Structure Processing Properties Relationships in Stoichiometric and Nonstoichiometric Oxides

Edited by Speranta Tanasescu

Published in London, United Kingdom

IntechOpen

Supporting open minds since 2005

Structure Processing Properties Relationships in Stoichiometric and Nonstoichiometric Oxides
http://dx.doi.org/10.5772/intechopen.77573
Edited by Speranta Tanasescu

Contributors
Netram Kaurav, Paras Dubey, S D Kaushik, A. K. Singh, Jaroslav Sestak, Paula Vilarinho, Speranta Tanasescu, Florina Teodorescu, Florentina Maxim, Alina Botea-Petcu, Ludwig J. Gauckler, Alexander Tkach

First published in London, United Kingdom, 2020 by IntechOpen
IntechOpen is the global imprint of INTECHOPEN LIMITED, registered in England and Wales, registration number: 11086078, 5 Princes Gate Court, London, SW7 2QJ, United Kingdom
Printed in Croatia

British Library Cataloguing-in-Publication Data
A catalogue record for this book is available from the British Library

Additional hard and PDF copies can be obtained from orders@intechopen.com

Structure Processing Properties Relationships in Stoichiometric and Nonstoichiometric Oxides
Edited by Speranta Tanasescu
p. cm.
Print ISBN 978-1-78985-451-0
Online ISBN 978-1-78985-452-7
eBook (PDF) ISBN 978-1-83969-130-0

For EU product safety concerns:
IN TECH d.o.o., Prolaz Marije Krucifikse Kozulić 3, 51000 Rijeka, Croatia, info@intechopen.com or visit our website at intechopen.com.

Meet the editor

Speranta Tanasescu, PhD, is Senior Researcher I and head of the Laboratory of Chemical Thermodynamics in the "Ilie Murgulescu" Institute of Physical Chemistry of the Romanian Academy, Bucharest. Dr. Tanasescu obtained a PhD in Physical Chemistry from the Romanian Academy in 1979 and has served as a supervisor in chemistry since 2002. She received the "Gh. Spacu" Award from the Romanian Academy in 1972. Her thematic research focuses on activities with impact in the following domains: materials science, nanoscience and nanotechnologies, new sources of energies, nanosafety, and nanomedicine. Her research is significant for understanding processing-structure relationships as well as for finding key parameters in relation to bio-reactivity of the nanomaterials with impact in both nanosafety and nanomedicine research.

Contents

Preface

The interrelation among composition, microstructure, and properties of stoichiometric and nonstoichiometric compounds is a major field of research for both scientific and technological reasons. This book focuses on metal oxides, which present a large diversity of electrical, magnetic, optical, optoelectronic, thermal, electrochemical, and catalytic properties, making them suitable for a wide range of applications. Searching for novel approaches and major breakthroughs in materials properties, researchers have conducted many studies on the synthesis, structure, and characterization of stoichiometric and nonstoichiometric (metal oxides) compounds. However, there are still questions regarding the relationships of all factors affecting the stability and reliability of these materials for specific applications.

Because a volume of this size could not cover all topics relevant for this major field of study, we have chosen in this book to concentrate on a limited number of chemical (oxide) systems that exemplify the complex bridging between material structure, synthesis, and properties.

An introductory chapter presents an up-to-date view on some important issues whose tight integration should provide a better understanding of the complex phenomena determining the fundamental properties of a large range of oxide systems. The chapter also defines the critical requirements for advances in design and application of next-generation materials.

Chapters describe various methods of synthesis (each of these methods leaving their own mark on the properties of the resulting materials) together with different advanced characterization methods. In addition, chapters present several advanced techniques for ascertaining stoichiometry. A special focus is on understanding the effect of conditions of synthesis on the degree of nonstoichiometry, as these aspects provide clues for controlling the evolution of properties. The book also emphasizes the role of thermodynamic parameters on the stabilization of the phase and physical properties in oxides.

It is hoped that this book brings more awareness about the significance of the relationship among structure, processing, and properties in stoichiometric and nonstoichiometric compounds.

As the editor, I would like to thank all the chapter authors for their contributions. Without them, it would not have been possible to create such a quality book that brings together scientific contributions with special emphasis on the interrelations between materials chemistry, processing, microstructures, and properties.

Dr. Speranta Tanasescu
Head of Labororatoy of Chemical Thermodynamics,
"Ilie Murgulescu" Institute of Physical Chemistry of the Romanian Academy,
Bucharest, Romania

Chapter 1

Introductory Chapter: Structure-Processing-Properties Relationships in Stoichiometric and Nonstoichiometric Oxides

Speranta Tanasescu

1. Introduction

The research in the area of stoichiometric and nonstoichiometric compounds has grown considerably in the past decades forming an emerging scientific issue with a great impact in the materials science. According to the law of definite proportions, stoichiometric compounds (also referred to as daltonides) are chemical compounds in which atoms are combined in exact whole-number ratios. By contrast, nonstoichiometric compounds (also known as berthollides) are chemical compounds deviating from stoichiometry, and therefore their elemental composition cannot be represented by a ratio of well-defined natural numbers [1–4]. The nonstoichiometry occurs most often in solids due to defects in the lattice of their crystalline structures, and it is most common in the transition metal oxides, but also the group of nonstoichiometric compounds includes nitrides, fluorides, hydrides, carbides, metal sulfides, tellurides, and so on [2, 4–8].

The focus of the present book is on metal oxides, which present a large diversity of electrical, magnetic, optical, optoelectronic, thermal, electrochemical, and catalytic properties, making them suitable for a wide range of applications including sensors, solid-state electronic devices, thermoelectric power generation, and energy harvesting. This richness of properties is owed to the oxides' structure flexibility (especially of transition metal oxides) that makes them easily distort/adapt to the relative sizes of the ions forming the compound [9]. This implies a large chemical diversity providing for a complex interplay of intrinsic materials properties (related to the constituent elements) and extrinsic (defect-driven) properties (related to the presence of impurities and/or dopants).

Recent developments in the solid-state chemistry motivated by the prospect of new applications with topics such as colossal magnetoresistance, multiferroics, high-entropy stabilization, and superconductivity have uncovered rich complexities [10–21] that had not previously been recognized. These studies bring up some important issues that should be taken into account when the development of new multifunctional materials was intended.

The core focus that is implicit in most of today's studies in this field is firmly on the fundamental understanding of the *materials chemistry and microstructures and how they are related to the thermodynamic, electronic, and transport properties of stoichiometric and nonstoichiometric compounds.*

Several basic textbooks address this topic in relation to the defect chemistry and nonstoichiometry, many of them emphasizing on oxides [22–26].

 IntechOpen

Nonstoichiometric crystals were first interpreted, as regards their structure and their thermodynamics, in terms of the statistical thermodynamics of point defects [1, 2]. At small deviations from the stoichiometric composition, an approach assuming noninteracting and randomly distributed point defects was used [24]. In order to describe the higher deviations from stoichiometry, the interactions between the defect complexes, clustering, and long-range ordering into superstructures had to be taken into account [3], and extended defects structure models were developed. Particularly, in variable valence transition metal oxides, there is a strong coupling between defect structures, charge ordering, and orbital degrees of freedom that impacts property evolution. In fact, the transition from the point defect concept in highly dilute defect systems to extended defects models applicable in highly defective systems reflect the need to better define important phenomena in the real world and to determine to a large extend, the fundamental properties of a large range of advanced materials used in heterogeneous catalysis, fuel cells, sensors, oxygen and hydrogen separation membranes, battery materials, electrochromics and so on.

Searching for novel approaches and major breakthroughs in the materials properties, important factors that influence materials stoichiometry, e.g., the variation of the defect structure under controlled conditions (various kinds of doping, temperature, oxygen pressure), were excellently addressed in a series of books, papers, and reviews, and several key physicochemical descriptors (measured as well as calculated ones) showing a good correlation between stoichiometry, structure, and properties have been described [27–35].

Providing insights into new possibilities to control and optimize the properties based on the *correlation between the thermochemical stability, the preparation routes, and characterization* of different oxide-based compositions is also a topic that underpins the development of emergent devices and technology.

Besides the classical synthesis approaches based on high-temperature synthesis (solid-state reaction, thermal decomposition, high-temperature/high-pressure preparation) or on electrochemical methods (anodic electrocrystallization, direct current electrolysis), new materials synthesis techniques have evolved, such as mechanosynthesis, microwave hydrothermal synthesis, and atomic layer deposition [4]. In addition, advances in materials synthesis techniques, such as molecular beam epitaxy, reflection high-energy electron diffraction (RHEED)-assisted growth, ion implantation, or nanopatterning of defects by focused ion beams, allow the production of materials with controlled concentrations of point or planar defects and create interstitial doping and local strain fields that can enable patterning of circuits and magnetic domains [13, 36–43].

One of the challenging problems related to the understanding and practical exploitation of the enhanced properties of nanocrystalline materials is the thermal stabilization of a nanoscale grain size. The thermal stability of these microstructures is essential for adopting nanocrystalline materials in commercial processes and applications. Because the refinement in grain size is accompanied by a significant increase in volume fraction of grain boundaries, the thermal stability involves not only the stability of the grain structure, i.e., microstructure, but also the stability of the structure of the grain boundaries in nanocrystalline materials [44]. The lowering of interfacial energy with grain refinement and lattice strain in nanometer-sized crystallites plays an important role in controlling grain size stability during the grain growth in nanocrystalline phases [45, 46].

In parallel with the development of the synthesis methods, the characterization of different oxides based compositions is a crucial issue. It was argued that, due to the particularities of nonstoichiometric compounds, only after a thorough analysis of the composition, structure, and properties one can conclude that the compound is a nonstoichiometric compound rather than a stoichiometric compound [4]. There is a complex task for which a combination of different methods is required. The results of the classical methods for the composition analysis, e.g., iodometry, cerimetry or electroanalysis methods, should be correlated with the crystallographic structure and microstructure information coming from application of X-ray diffraction techniques, neutron diffraction, high-resolution electron microscopy (HRTEM), laser Raman spectroscopy (LRS) and electron paramagnetic resonance (EPR), together other advanced technique imaging at the atomic level and allowing a detailed study of local defect structures and chemistry, e.g., scanning transmission electron microscopy (STEM) and in situ electron energy loss spectroscopy (EELS) [47–52]. In addition, many advances have come by measuring the physical properties such as electrical conductivity, magnetic, optical, and optoelectronic properties that strongly depend on stoichiometry and on types and concentration of the defects.

The oxide microstructure modification by using different synthetic methods and the modification of various compositional variables such as the nature and concentration of donor- or acceptor-type dopants are essential for obtaining optimum electrical and transport characteristics. Heat treatment is also an important step not only to ensure stability but also to control structural defects and grain size, also contributing to sensitivity and selectivity of the new materials. Previous reports on the substituted perovskites indicate that the mismatch at the A and B sites in the ABO_3 structure creates strain on grain boundaries which affect not only the electrical but also the thermodynamic properties [9], phase stability, and oxygen stoichiometry [53, 54]. It was also pointed out that the remarkable behavior of the multiferroic and magnetoresistive materials, as well as of the mixed ionic-electronic conducting ceramic membranes obtained by substitution of A and B sites, could be explained not only qualitatively by the structural changes upon doping but also by the fact that the thermodynamic properties are extremely sensitive to the chemical defects in oxygen sites [55–57]. An interesting relationship between the energetics of growth film conditions and the subsequent materials properties was observed when the pulsed laser deposition (PLD) was used in the synthesis of complex oxide films. Variations in the energetics of growth process can enable fine-tuning and control stoichiometry, dielectric response, thermal and electrical conductivity of films and heterointerfaces [58, 59].

The role of the energetic parameters in understanding the physical- and chemical-modified properties associated with the rise of the surface/volume ratio at a nanometer scale is also a topic of paramount importance [60]. Shifts in thermodynamics at the nanoscale and the strong interplay between the thermodynamic properties and electrical and structural characteristics in the hydrothermally prepared perovskite materials have been revealed [61]. In addition, at the nanometric scale, a large variety of morphologies and related surface properties can exist for the same metal oxide. This means that a great deal of attention must be turned to the energetic parameters which play an important role in the overall properties and behavior of materials. Exploring the relationships between morphology and thermodynamic properties of nanocrystalline $BaTiO_3$, it was shown that the enhancement of the dielectric properties for the $BaTiO_3$ hydrothermal-prepared powders with 1D morphology, comparatively with nanocubes or hollow-type morphologies, is strongly correlated with the increase in the binding energy of oxygen in the perovskite structure [62, 63].

Computational approaches, such as DFT-based calculations, phase-field modeling, molecular dynamics (MD) and Monte Carlo simulations) integrated at different stages of materials development have demonstrated to be important tools to address the complexity of based oxides stoichiometric and nonstoichiometric compounds and provide information on phase competition and stability, defect dynamics and kinetics and so on. Thermodynamic Databases, such as CALPHAD (CALculation of PHAse Diagrams) or Databases including high-throughput DFT calculations, read-across and QSAR approaches, together with machine learning platforms have been developed contributing to the prioritization and screening of materials properties for applications as electronics, fuel cells, multiferroics, piezoelectrics, magnetocalorics, thermoelectrics [51, 64–73].

Critical advances in discovery and design of next-generation materials are expected by applying *the concept of the Materials Genome Initiative (MGI)* [74, 75] that tightly integrates high-throughput experiment (including both synthesis and characterization), theory, and computation. The development of advanced materials by using emerging synthetic and processing approaches should be based on the understanding of all factors affecting the reliability of these materials for specific applications (including a thorough thermodynamic analysis). The development of robust experimental methods to quantify microstructure and interfaces and to identify descriptors that strongly correlate with rearrangement dynamics on multiple lengthscales is also necessary. Multiscale and lifetime modeling should be accompanied by designing new tools and data analytics enabling high-content analysis and automated data evaluation and thus increasing the ability to understand and tailor the physical properties of materials. The strong interplay between these components offers great opportunities to establish support at the different tiered workflows directed toward emergent applications.

Driven by this concept, we have chosen in this book to concentrate on a limited number of chemical systems that exemplify the complex bridging between materials structure, synthesis, and properties. A special focus is on the role of thermodynamic parameters on the stabilization of the phase and physical properties in oxides. Various methods of synthesis are employed, each of these methods leaving their own mark on the properties of the resulting materials. The strong structure-processing-property relationship is emphasized in each of the chapters of this book, as can be seen from the brief overview of the main topics developed in these chapters: (i) The synthesis and complex characterization of a transitional metal oxide extensively used in industry, e.g., nickel oxide, are discussed. The understanding of the conditions of synthesis effect on the degree of nonstoichiometry provides clues for controlling the properties evolution. (ii) The stoichiometry and nonstoichiometry from crystal structure point of view are introduced along with some examples relevant for the importance of nonstoichiometry in the application - oriented research. Several advanced techniques available to ascertain stoichiometry are presented with a special emphasis on neutron diffraction techniques. Finally, important results obtained using neutron diffraction and scattering in identifying the structural modification which leads to superconductivity in the compounds are described. (iii) Particular aspects of the thermodynamic concepts related to associated phase equilibria in oxides exhibiting variable stoichiometry are emphasized. Insights into the equilibrium studies and construction of thermodynamic models of nonstoichiometric phases with application in high temperature superconducting materials are providing. (iv) The scientific and technological importance of the stoichiometry variation in the lead-free perovskite-structure materials, such as $SrTiO_3$ (ST) and $KTaO_3$ (KT), pure or modified, are defined. The strong relationship between the grain growth, the Sr/Ti or K/Ta ratio, the phase structure, morphology and dielectric response of ST and KT ceramics is overviewed. (v) The strong correlation between structure, nonstoichiometry and thermodynamic properties of some mixed conducting perovskite-type oxides $Ba_xSr_{1-x}Co_{1-y}Fe_yO_{3-\delta}$ (BSCF) studied as potential high-performance solid oxide fuel cells cathode materials is discussed and the effect of A- and B-site dopants concentration and of the oxygen stoichiometry change on the thermodynamic stability and morphology of the BSCF samples was evidenced.

We hope the approach adopted on this book would give an account about the significance of the structure-processing-property relationship in stoichiometric and nonstoichiometric compounds as an important issue for both scientific and applicative reasons.

Author details

Speranta Tanasescu
"Ilie Murgulescu" Institute of Physical Chemistry of the Romanian Academy, Bucharest, Romania

*Address all correspondence to: stanasescu2004@yahoo.com

References

[1] Anderson JS. Defect chemistry and non-stoichiometric compounds. In: Rao CNR, editor. Modern Aspects of Solid State Chemistry. Boston, MA: Springer; 1970. DOI: 10.1007/978-1-4684-1875-0_3

[2] Anderson JS. Nonstoichiometric compounds: A critique of current structural views. Proceedings - Indian Academy of Sciences, Chemical Sciences. 1984;**93**:861-904. DOI: 10.1007/BF02840335

[3] Collongues R. Nonstoichiometry in oxides. Progress in Crystal Growth and Characterization of Materials. 1992;**25**(4):203-240. DOI: 10.1016/0960-8974(92)90013-G

[4] Zhang JL, Hong GY. Nonstoichiometric compounds. In: Ruren X, Yan X, editors. Modern Inorganic Synthetic Chemistry. 2nd ed. Elsevier; 2017. pp. 329-354. DOI: 10.1016/B978-0-444-63591-4.00013-6

[5] Rao CNR. Transition metal oxides. Annual Review of Physical Chemistry. 1989;**40**:291-326. DOI: 10.1146/annurev.pc.40.100189.001451

[6] Lipatnikov VN, Zueva LV, Gusev AI, Kottar A. Disorder-order phase transformations and electrical resistivity of nonstoichiometric titanium carbide. Physics of the Solid State. 1998;**40**:1211-1218. DOI: 10.1134/1.1130523

[7] Gusev AI. Order-disorder transformations and phase equilibria in strongly nonstoichiometric compounds. Uspekhi Fizicheskikh Nauk. 2000;**170**:3-37. DOI: 10.1070/PU2000v043n01ABEH000647

[8] Rempel AA. Nonstoichiometric transition metal compounds. A review. In: Proceedings of the 17th Israeli-Russian Bi-National Workshop; 2018. pp. 167-189

[9] Stølen S, Bakken E, Mohn CE. Oxygen-deficient perovskites: Linking structure, energetics and ion Transport. Physical Chemistry Chemical Physics. 2006;**8**:429-447. DOI: 10.1039/b512271f

[10] Coey JMD, Viret M, von Molnar S. Magnetoresistance effect. Advances of Physics. 1999;**48**:167. DOI: 10.1080/00018730903363184

[11] Dabrowski B, Kolesnik S, Chmaissem O, Bukowski Z, Mais J, Kimball CW, et al. Spectacular magneto-related properties of complex oxides. In: Gibbs MRJ, editor. Modern Trends in Magnetostriction Study and Application. NATO Science Series (Series II: Mathematics, Physics and Chemistry), Vol. 5. Dordrecht: Springer; 2001. DOI: 10.1007/978-94-010-0959-1_10

[12] Tokura Y. Critical features of colossal magnetoresistive manganites. Reports on Progress in Physics. 2006;**69**:797-851. DOI: 10.1088/0034-4885/69/3/R06

[13] Mundy J, Brooks C, Holtz M, Moyer Jarrett A, Das H, Rébola AF, et al. Atomically engineered ferroic layers yield a room-temperature magneto electric multiferroic. Nature. 2016;**537**:523-527. DOI: 10.1038/nature19343

[14] Rost CM, Sachet E, Borman T, Moballegh A, Dickey EC, Hou D, et al. Entropy-stabilized oxides. Nature Communications. 2015;**6**:8485. DOI: 10.1038/ncomms9485

[15] Bérardan D, Franger S, Dragoe D, Meena AK, Dragoe N. Colossal dielectric constant in high entropy oxides. Physica Status Solidi. 2016;**10**(4):328-333. DOI: 10.1002/pssr.201600043

[16] Berardan D, Meena AK, Franger S, Herrero C, Dragoe N. Controlled

Jahn-Teller distortion in (MgCoNiCuZn) O-based high entropy oxides. Journal of Alloys and Compounds. 2017;**704**:693-700

[17] Jaing S, Hu T, Gild J, Zhou N, Nie J, Qin M, et al. A new class of high-entropy perovskite oxides. Scripta Materialia. 2018;**142**:116-120. DOI: 10.1016/j.scriptamat.2017.08.040

[18] Dragoe N, Bérardan D. Entropy stabilization provides a new direction for developing functional materials. Science. 2019;**366**(6465):573-574. DOI: 10.1126/science.aaz1598rapid.com

[19] Osenciat N, Bérardan D, Dragoe D, Leridon B, Holé S, Meena AK, et al. Charge compensation mechanisms in Li-substituted high-entropy oxides and influence on Li superionic conductivity. Journal of the American Ceramic Society. 2019;**102**:6156-6162. DOI: 10.1111/jace.16511

[20] Matin MA, Hossain MN, Ali MA, Hakim MA, Islam MF. Enhanced dielectric properties of prospective Bi0.85Gd0.15Fe_1-xCrxO_3 multiferroics. Results in Physics. 2019;**12**:1653-1659 DOI: 10.1016/j.rinp.2019.01.079

[21] Witte R, Sarkar A, Kruk R, Eggert B, Brand RA, Wende H, et al. High-entropy oxides: An emerging prospect for magnetic rare-earth transition metal perovskites. Physical Review Materials. 2019;**3**:034406

[22] Wagner C. Über den Mechanismus der elektrischen Stromleitung in Nernststift. Die Naturwissenschaften. 1943;**31**:265-268

[23] Kofstad P. Nonstoichiometry, Diffusion and Electrical Conductivity in Binary Metal Oxides. New York, London, Sydney, Toronto: Wiley-Interscience; 1972. DOI: 10.1002/maco.19740251027

[24] Kröger FA. Chemistry of Imperfect Crystals. Amsterdam: North-Holland Publishing Company; 1964. (OCoLC) 610606600

[25] Sørensen OT, editor. Non-stoichiometric Oxides. Academic Press; 1981. p. 441. DOI: 10.1016/B978-0-12-655280-5. X5001-9

[26] Kosuge K. Chemistry of Non-stoichiometric Compounds. USA: Oxford Press; 1994. DOI: 10.1021/ja945020d

[27] Centi G. Role of non-stoichiometry and soft chemistry in the preparation of advanced catalysts. Defect and Diffusion Forum. 2001;**191**:17-34. DOI: 10.4028/www.scientific.net/DDF.191.17

[28] Søgaard M, Hendriksen PV, Poulsen FW. Determination of transport and catalytic properties of mixed ionic and electronic conductors using transient responses. In: Linderoth S et al., editors. Solid State Electrochemistry. Proceedings of the Risø International Symposium on Materials Science; Risø (DK); 4-8 September 2005. Roskilde: Risø National Laboratory; 2005. pp. 355-362

[29] Hendriksen PV, Larsen PH, Mogensen M, Poulsen FW, Wiik K. Prospects and problems of dense oxygen permeable membranes. Catalysis Today. 2000;**56**:283-295. DOI: 10.1016/S0920-5861(99)00286-2

[30] Hashimoto SI, Kammer Hansen K, Larsen PH, Poulsen FW, Mogensen M. A study of $Pr_{0.7}Sr_{0.3}Fe_{1-x}Ni_xO_{3-\delta}$ as a cathode material for SOFCs with intermediate operating temperature. Solid State Ionics. 2005;**176**:1013-1020. DOI: 10.1016/j.ssi.2004.09.010

[31] Sun G, Thygesen A, Ale MT, Mensah M, Poulsen FW, Meyer AS. The significance of the initiation process parameters and reactor design for maximizing the efficiency of microbial fuel cells. Applied Microbiology and Biotechnology. 2014;**98**:2415-2427. DOI: 10.1007/s00253-013-5486-5

[32] Ovtar S, Søgaard M, Norrman K, Hendriksen PV. Oxygen Exchange and Transport in $(La_{0.6}Sr_{0.4})_{0.98}FeO_{3-d}$–$Ce_{0.9}Gd_{0.1}O_{1.95}$ Dual-Phase Composites. Journal of the Electrochemical Society. 2018;**165**(3):F220-F231. DOI: 10.1149/2.1031803jes

[33] Fergus JW. Oxide materials for high temperature thermoelectric energy conversion. Journal of the European Ceramic Society. 2012;**32**:525-540. DOI: 10.1016/j.jeurceramsoc.2011.10.007

[34] Somani PR, Radhakrishnan S. Electrochromic materials and devices: Present and future. Materials Chemistry and Physics. 2002;**77**:117-133. DOI: 10.1016/S0254-0584(01)00575-2

[35] Llordés A, Wang Y, Fernandez-Martinez A, Xiao P, Lee T, Poulain A, et al. Linear topology in amorphous metal oxide electrochromic networks obtained via low-temperature solution processing. Nature Materials. 2016;**15**:1267-1273. DOI: 10.1038/nmat4734

[36] Xu RR, Pang WQ. Inorganic Synthesis and Preparation Chemistry. Beijing: Higher Education Press; 2001. p. 44

[37] Takamura Y, Chopdekar RV, Scholl A, Doran A, Liddle JA, Harteneck B, et al. Tuning magnetic domain structure in nanoscale $La_{0.7}Sr_{0.3}MnO_3$ islands. Nano Letters. 2006;**6**(6):1287-1291. DOI: 10.1021/nl060615f

[38] Lee C, Orloff N, Birol T, Zhu Y, Goian V, Rocas E, et al. Exploiting dimensionality and defect mitigation to create tunable microwave dielectrics. Nature. 2013;**502**:532-536. DOI: 10.1038/nature12582

[39] Mathew S, Annadi A, Chan TK, Asmara TC, Zhan D, Wang XR, et al. Tuning the interface conductivity of $LaAlO_3/SrTiO_3$ using ion beams: Implications for patterning. ACS Nano. 2013;**7**(12):10572-10581. DOI: 10.1021/nn4028135

[40] Yang Y, Liu XL, Yang Y, Xiao W, Li ZW, Xue DS, et al. Synthesis of nonstoichiometric zinc ferrite nanoparticles with extraordinary room temperature magnetism and their diverse applications. Journal of Materials Chemistry C. 2013;**1**(16):2875-2885. DOI: 10.1039/C3TC00790A

[41] Saremi S, Dedon LR, Mundy JA, Hsu SL, Chen Z, Damodaran AR, et al. Enhanced electrical resistivity and properties via ion bombardment of ferroelectric thin films. Advanced Materials. 2016;**28**:10750-10756. DOI: 10.1002/adma.201603968

[42] Damodaran AR, Breckenfeld E, Chen Z, Lee S, Martin LW. Enhancement of ferroelectric Curie temperature in $BaTiO_3$ films via strain induced defect dipole alignment. Advanced Materials. 2014;**26**:6341-6347. DOI: 10.1002/adma.201400254

[43] Saremi S, Gao R, Dasgupta A, Martin LW. New facets for the role of defects in ceramics. American Ceramic Society Bulletin. 2018;**97**(1):16-23

[44] Koch CC. Processing structure property relationships in ultrafine grain and nanocrystalline materials. JPCS. 2009;**144**:1-8. DOI: 10.1088/1742-6596/144/1/012081

[45] Murty BS, Datta MK, Pabi SK. Structure and thermal stability of nanocrystalline materials. Sadhana. 2003;**28**:23-45. DOI: 10.1007/BF02717124

[46] Gleiter H. Nanostructured materials: Basic concepts and microstructure. Acta Materialia. 2000;**48**:1-29. DOI: 10.1016/S1359-6454(99)00285-2

[47] Xu W, Bowes PC, Grimley ED, Irving DL, Le Beau JM. In-situ realspace

imaging of single crystal surface reconstructions via electron microscopy. Applied Physics Letters. 2016;**109**:201601. DOI: 10.1063/1.4967978

[48] Seebauer EG, Kratzer MC. Experimental and computational characterization. In: Charged Semiconductor Defects: Structure, Thermodynamics and Diffusion. London: Springer; 2009. pp. 39-61. DOI: 10.1007/978-1-84882-059-3

[49] Karppinen M, Yamauchi H. Chemical design of copper-oxide superconductors: Homologous series and oxygen engineering in frontiers. In: Narlikar AV, editor. Superconducting Materials. Berlin: Springer Verlag; 2005. pp. 255-294

[50] Yadav AK, Nelson CT, Hsu SL, Hong Z, Clarkson JD, Schlepüetz CM, et al. Observation of polar vortices in oxide superlattices. Nature. 2016;**530**:198-201. DOI: 10.1038/nature16463

[51] Bartel CJ, Sutton C, Goldsmith BR, Ouyang R, Musgrave CB, Ghiringhelli LM, et al. New tolerance factor to predict the stability of perovskite oxides and halides. Science Advances. 2019;**5**(2):eaav0693. DOI: 10.1126/sciadv.aav0693

[52] Itahashi YM, Ideue T, Saito Y, Shimizu S, Ouchi T, Nojima T, et al. Nonreciprocal transport in gate-induced polar superconductor $SrTiO_3$. Science Advances. 2020;**6**(13):eaay9120. DOI: 10.1126/sciadv.aay9120

[53] McIntosh S, Vente JF, Haije WG, Blank DHA, Bouwmeester HJM. Phase stability and oxygen non-stoichiometry of SrCo0.8Fe0.2O3−δ measured by in situ neutron diffraction. Solid State Ionics. 2006;**177**:833-842. DOI: 10.1016/j.ssi.2006.02.017

[54] Kuhn M, Kim JJ, Bishop SR, Tuller HL. Oxygen nonstoichiometry and defect chemistry of perovskite-structured $Ba_xSr_{1-x}Ti_{1-y}Fe_yO_{3-y/2+\delta}$ solid solutions. Chemistry of Materials. 2013;**25**:2970-2975. DOI: 10.1021/cm400546z

[55] Tanasescu S, Petcu A, Ianculescu A. Chapter 15: Effects of doping and nonstoichiometry on the thermodynamic properties of multiferroic ceramics. In: Lallart M, editor. Ferroelectrics—Vol. II. Rijeka, Croatia: InTech; 2011. pp. 347-372. DOI: 10.5772/16392

[56] Tanasescu S, Maxim F, Teodorescu F, Giurgiu LM. Influence of composition and particle size on spin dynamics and thermodynamic properties of magnetoresistive perovskites. Journal of Nanoscience and Nanotechnology. 2008;**8**(2):914-923. DOI: 10.1166/jnn.2008.D127

[57] Tanasescu S, Yáng Z, Martynczuk J, Teodorescu F, Botea A, Totir N, et al. Effects of A-site composition and oxygen nonstoichiometry on the thermodynamic stability of compounds in the Ba-Sr-Co-Fe-O system. Solid State Chemistry. 2013;**200**:354-362. DOI: 10.1016/j.jssc.2013.01.030

[58] Breckenfeld E, Wilson R, Karthik J, Damodaran AR, Cahill DG, Martin LW. Effect of growth induced (non)stoichiometry on the structure, dielectric response, and thermal conductivity of $SrTiO_3$ thin films. Chemistry of Materials. 2012;**24**:331. DOI: 10.1021/cm203042q

[59] Breckenfeld E, Bronn N, Mason N, Martin LW. Tunability of conduction at the $LaAlO_3/SrTiO_3$ heterointerface: Thickness and compositional studies. Applied Physics Letters. 2014;**105**:121610. DOI: 10.1063/1.4896778

[60] Navrotsky A, Ma C, Lilova K, Birkner N. Nanophase transition metal oxides show large thermodynamically

driven shifts in oxidation-reduction equilibria. Science. 2010;**330**:199. DOI: 10.1126/science.1195875

[61] Rusti CF, Badilita V, Sofronia AM, Taloi D, Anghel EM, Maxim F, et al. Thermodynamic properties of $Ba_{0.75}Sr_{0.25}TiO_3$ nanopowders obtained by hydrothermal synthesis. Journal of Alloys and Compounds. 2017;**693**:1000-1010. DOI: 10.1016/j.jallcom.2016.09.215

[62] Maxim F, Poenaru I, Teodorescu F, Tanasescu S. Barium titanate torus-like particles: Low-temperature synthesis and formation mechanism. European Journal of Inorganic Chemistry. 2014;**30**:5160-5167. DOI: 10.1002/ejic.201402497

[63] Maxim F, Berger D, Teodorescu F, Hornoiu C, Lete C, Tanasescu T. Low-temperature synthesis and thermodynamic and electrical properties of barium Titanate Nanorods. Journal of Nanomaterials. 2015;**1**:1-10. DOI: 10.1155/2015/827641

[64] Olson GB. Designing a new material world. Science. 2000;**288**:993-998

[65] Greeley J, Jaramillo T, Bonde J, Chorkendorff I, Nørskov JK. Computational high-throughput screening of electrocatalytic materials for hydrogen evolution. Nature Materials. 2006;**5**:909-913. DOI: 10.1038/nmat1752

[66] Poulsen FW. Methods and limitations in defect chemistry modelling [Thesis]. Roskilde, Denmark: Risø National Laboratory, DTU; 2007

[67] National Research Council. Integrated Computational Materials Engineering: A Transformational Discipline for Improved Competitiveness and National Security. Washington, DC: The National Academies Press; 2008. p. 152

[68] Kattner UR, Campbell CE. Invited review: Modelling of thermodynamics and diffusion in multicomponent systems. Materials Science and Technology. 2009;**25**:25443-25459. DOI: 10.1179/17432840837200

[69] Roy A, Benntt JW, Rabe K, Vanderbilt D. Half-heusler semiconductors as piezoelectrics. Physical Review Letters. 2012;**109**(3):037602. DOI: 10.1103/PhysRevLett.109.037602

[70] Curtarolo S, Hart GL, Nardelli MB, Mingo N, Sanvito S, Levy O. The high-throughput high way to computational materials design. Nature Materials. 2013;**12**(3):191-201. DOI: 10.1038/nmat3568

[71] Marzari N. Materials modelling: The frontiers and the challenges. Nature Materials. 2016;**15**:381-382. DOI: 10.1038/nmat4613

[72] Montoya J, Seitz L, Chakthranont P, Vojvodic A, Jaramillo TF, Nørskov JK. Materials for solar fuels and chemicals. Nature Materials. 2016;**16**:70-81. DOI: 10.1038/nmat4778

[73] Jain A, Shin Y, Persson KA. Computational predictions of energy materials using density functional theory. Nature Reviews Materials. 2016;**1**:15004. DOI: 10.1038/natrevmats.2015.4

[74] de Pablo JJ, Jones B, Kovacs CL, Ramirez AP. The materials genome initiative, the interplay of experiment, theory and computation. Current Opinion in Solid State & Materials Science. 2014;**18**:99-117. DOI: 10.1016/j.cossms.2014.02.003

[75] de Pablo JJ, Jackson NE, Webb MA, Chen LQ, Moore JE, Morgan D, et al. New frontiers for the materials genome initiative. Computational Materials. 2019;**5**:41. DOI: 10.1038/s41524-019-0173-4

Chapter 2

Role of Neutron Diffraction in Identifying Stoichiometry and Nonstoichiometry in the Compounds

Som Datta Kaushik and Anil Kumar Singh

Abstract

In this chapter we introduce stoichiometry and nonstoichiometry from crystal structure point of view along with some examples. We also discussed about the importance of nonstoichiometry in the application oriented research work and their use in the technological applications. We further discuss the ways to identify stoichiometry through various methods. We then introduce neutron diffraction and briefly describe how neutrons and X-ray interacts with matter and the difference in their interaction with matter. We then focus upon its (neutron) usability to identify nonstoichiometry by using some examples available in the literatures. High-temperature superconductivity-based research has seen the importance of neutron diffraction and scattering in identifying the structural modification which leads to superconductivity in the compounds.

Keywords: neutron, interaction with matter, neutron diffraction, neutron scattering, oxygen deficiency

1. Introduction

Stoichiometry in plane Google search is defined as "the relationship between the relative quantities of substances taking part in a reaction or forming a compound, typically a ratio of whole integers." This points out that stoichiometry has something to do with the formation of a compound. Thus in order to define the stoichiometric compound, we can consider the compound in which the reactant or constituent elements of the compound are in integer numbers. It is usually proposed that stoichiometry follows the law of conservation of mass where the total mass of the reactant element is equal to the total mass of the product, thus leading us to believe that the relations among quantities of reactants and products typically form a ratio of positive integers. This means that if the amounts of the separate reactants are known, then the amount of the product can be calculated. If we take the example of water, the chemical formula of which is H_2O, thus the total mass of water molecule is the sum of the mass of two hydrogen atoms and one oxygen atom. Here, two molecules of hydrogen react with one molecule of oxygen gas to yield two molecules of water. This particular chemical equation is an example of complete reaction. Stoichiometry measures these quantitative relationships and is used to determine the amount of products and reactants that are produced or needed in a given reaction. Defining quantitative relationships among substances as they participate in chemical reactions is known as reaction stoichiometry. In the example above, reaction stoichiometry measures the relationship between hydrogen and oxygen as they react to water.

We can further define that stoichiometric compounds are those compounds which maintain their stoichiometry or simply maintain the ratio in which they were formed. Here we must understand that stoichiometric compounds can still possess the defects which are termed as intrinsic point defects. As displayed in **Figure 1**, some of the well-known defects are called Schottky (originating due to cation and anion vacancies), Anti-Schottky (due to cation and anion interstitials), Frankel (cation vacancies and interstitials), Anti-Frenkel (anion vacancies and interstitials), and Anti-site defects (originating because of cation and anion swap).

Upon the aforementioned analogy, we can describe the nonstoichiometric compound as antithesis of stoichiometric compounds. Thus we can describe nonstoichiometric compounds as chemical compounds, having elemental composition whose proportions cannot be represented by integers; most often, in such materials, some small percentage of atoms are missing, or too many atoms are packed into an otherwise perfect lattice work. In order to define nonstoichiometry microscopically, a situation of array of atom (shown by the open circle in **Figure 1**) in two-dimensional primitive cubic crystal can be imagined. In such scenario the absence of the atom from the regular place will form vacancy (described as I), displacement of the atom will create Frenkel pair (described as II), and substitution of the atom by another smaller or larger atom (shown by closed circle) can be referred as nonstoichiometry. Nikolai Semenovich Kurnakov along with his students has worked quite significantly on stoichiometry. Kurnakov et al. suggested that the nonstoichiometric compound could be allocated in the categories of berthollides and daltonides depending upon the monotonicity in property and composition correlation. The term berthollide was recognized

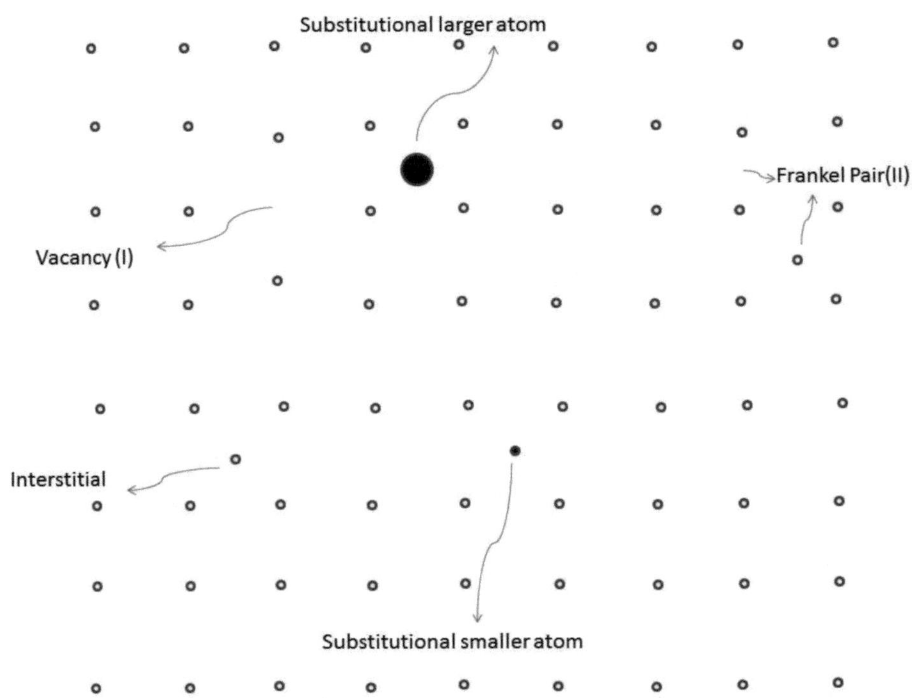

Figure 1.
A two-dimensional arrangement of atoms to describe nonstoichiometry.

by the International Union of Pure and Applied Chemistry (IUPAC) [1]. The name "berthollide" was formed from combination of two well-known chemists of nineteenth century, Claude Louis Berthollet and John Dalton, who proposed the substance composition theories, albeit these theories were in opposite direction and later Dalton's theory (Law of definite proportions) prevailed with significant exceptions [2].

2. Example of stoichiometric and nonstoichiometric materials and applications

Though it is generally said that nature loves symmetry, it can also be mentioned that stoichiometry is more like a concept in chemistry. In the real world, the naturally occurring stoichiometric materials are not often found. There are few examples of stoichiometric compounds including NaCl, Mgo, Al_2O_3, etc. Most of stoichiometry compounds are man-made, while on the other hand in nonstoichiometric compounds, in order to maintain the overall electrical neutrality, the defects are compensated by alternation in the valence state of other atoms present in the solid or by replacing the atom with atom of some other elements with a suitable charge. Many oxides and sulfides of metals are examples of nonstoichiometric compounds; here a rarely found stoichiometric iron(II) oxide, having the formula FeO, is an example of stoichiometric compound; however the more common material is nonstoichiometric compound, having the formula $Fe_{0.95}O$. Here one should note that, in nonstoichiometric compounds, the type of equilibrium defects can vary with attendant variation, thus resulting in the variation in bulk properties of the material [3]. Sulfides of iron also known as pyrrhotite are also examples of a nonstoichiometric inorganic compound. They have a general formula $Fe_{1-x}S$ ($x = 0$ to 0.2). The monosulfides of the transition metals (or better known as 3d elements) are often nonstoichiometric. The rare stoichiometric FeS end-member is known as the mineral troilite. Pyrrhotite is remarkable in that it has number of poly types, i.e., crystalline forms differing in symmetry (monoclinic or hexagonal) and composition (Fe_7S_8, Fe_9S_{10}, $Fe_{11}S_{12}$, and others). Most of the time, the aforementioned materials possess iron deficiency owing to the existence of lattice defects, mainly iron vacancies. In spite of the presence of these defects, the composition of these materials is normally expressed as a ratio of large numbers, and the crystals symmetry is reasonably high. This situation suggests that the iron vacancies in the materials are not randomly scattered over the crystal; rather it forms certain regular configurations. The vacancies significantly modify the magnetic properties of pyrrhotite: the magnetism gets enhanced with the concentration of vacancies and is absent for the stoichiometric FeS [4].

The presence of defects in nonstoichiometric compounds leads to display distinctive electrical or chemical properties; for example, when certain atoms are missing, electrons can travel through the solid more rapidly. Nonstoichiometric compounds are very useful in day-to-day life and found applications in various properties, viz., in ceramic and superconducting material, electrochemical (i.e., battery) system designs, and catalysis-based processing. Various valuable compounds can be produced by the reactions of hydrocarbons with oxygen in which a certain metal oxide acts as catalysis. Under this process transfer of "lattice" oxygen to the hydrocarbon substrate takes place, which temporarily generates a vacancy (or defect,) and subsequently, the missing oxygen is replenished by O_2. Such catalysts depend upon the ability of the metal oxide to form phases that are not stoichiometric [5]. Similar sequence of events describes other kinds of atom transfer reactions such as hydrogenation and hydride sulfurization catalyzed by solid catalysts. These considerations point out that stoichiometry is determined by the interior of crystals: the surfaces of crystals mostly do not follow stoichiometry of the bulk. The complex structures on surfaces are termed as "surface reconstruction."

A very interesting aspect of nonstoichiometric compound has been seen in high-temperature superconductors (HTSC). Many such HTSC's are nonstoichiometric. For example, yttrium barium copper oxide (YBCO) arguably is the most notable high-temperature superconductor and is a nonstoichiometric solid with the formula $YxBa_2Cu_3O_{7-x}$. The critical temperature (often denoted as Tc, which determines the temperature under which compounds become superconductor) of the superconductor depends on the exact value of x. The stoichiometric species has the value of x as zero, but this value can be as high as one. In the subsequent section, we will also deal with such examples. Here it is interesting to note that the identification of stoichiometry/nonstoichiometry in itself is an interesting science.

3. Techniques to characterize stoichiometry and nonstoichiometry

In order to identify stoichiometry, one has to carefully identify the ratio of constituent elements. Previously wet chemical method was employed to identify stoichiometry, but nowadays several advanced techniques are available to ascertain stoichiometry or otherwise. Some of the prominent techniques include high-resolution transmission electron microscope (HRTEM), field emission surface electron microscope (FESEM), energy-dispersive X-ray spectroscopy (EDS), etc. Here we will discuss neutron diffraction, which broadly assists in identifying various types of stoichiometry in terms of oxygen deficiencies, anti-site disorder, and cation distribution. There are methods which can be used to estimate stoichiometry by studying the crystal structure as well. X-ray diffraction and neutron diffraction are two such methods. Here neutron diffraction is a more powerful technique than X-ray diffraction. We will try to understand it in terms of various examples explained in subsequent sections. Here before going further, it would be interesting to understand how the neutron interacts with the matter.

4. X-ray and neutron diffraction techniques and neutron interactions with matter

X-ray diffraction (XRD) has been a very popular technique to identify the crystal structure due to more accessibility in terms of various parameters such as economics, portability, and hassle-free operation and thus has more prominently been described in the materials science books than neutron diffraction. Basic principles of X-ray as well as neutron scattering and neutron diffraction along with its usage in characterization technique of materials have been the subject matter of several excellent books which can be consulted for details [6–15]. Broadly neutrons interaction with matter can be defined in variety of ways due to which neutron diffraction seems similar to and but different from X-ray diffraction. First of all a neutron is a chargeless particle. The chargeless state of neutron assists to penetrate deep into the material, approximately up to the atomic nuclei which looks like a point in the whole material. In this scenario nuclear scattering takes place at very short distances of close to 10^{-14}–10^{-15} m; due to this scattering form, an ordered crystalline material yields coherent Bragg scattering. This Bragg scattering produces the intensity in the form of various Bragg's reflections corresponding either in single-crystal or in the powder diffraction pattern. These results are more or less similar to X-ray diffraction and provide essentially the same structural information. However, there is a point to be noted that the elemental scattering lengths for X-ray and neutron are distinctly different. The reason lies in the fact that X-ray diffraction results from scattering by the electrons; thus atoms of nearby atomic number exhibit very similar scattering which make practically difficult to distinguish between the nearby atoms by X-ray diffraction. On the other hand, neutron scattering depends upon the nature of the scattering nuclide, and consequently, atoms of similar atomic number often have quite different neutron scattering lengths (**Figure 2**).

The difference in scattering length is also apparent in **Table 1**, for some of the important transition metals. In addition to this, it is also to be noted that for neutron scattering the integrated intensity is independent of scattering angle (θ), while for the X-ray scattering, it is highly dependent upon the scattering angle. This difference can be observed in the angular dependence of the form factor for the different scattering processes (**Figure 3**).

This can be observed in the peak intensity for X-ray diffraction and neutron diffraction pattern. For X-ray diffraction pattern peak, intensity gradually decreases as scattering angle increases, while for the neutron diffraction, it is not the case. This situation helps in extracting the comprehensive information post analysis of neutron diffraction data.

The major factor is the fact that the neutron possesses the magnetic moment, because of this, the neutron is also scattered as a result of interaction with magnetic ion (in ordered form) present within a material; thus in the resultant coherent scattering, the magnetic scattering is superimposed upon the nuclear scattering in such magnetic material, though this magnetic scattering may not certainly affect the orientation of spin of the neutron, but since the magnetic neutron scattering consists of interaction with spin and orbital components of the magnetic moments, thus resultant moments originate from all the unpaired valence electrons. Here one should note that the magnetic scattering has an angular dependence which is similar to that of X-ray scattering by electrons as shown in **Figure 3**. This situation is opposite to the nuclear scattering, where the nuclear scattering is independent of scattering angle as displayed in **Figure 3** in which the typical angular dependence of the spin, orbital, and nuclear form factors for neutron scattering by chromium is shown, and for comparison, the form factor for X-ray scattering is also depicted. In order to identify the magnetic and nuclear neutron scattering, angular dependence

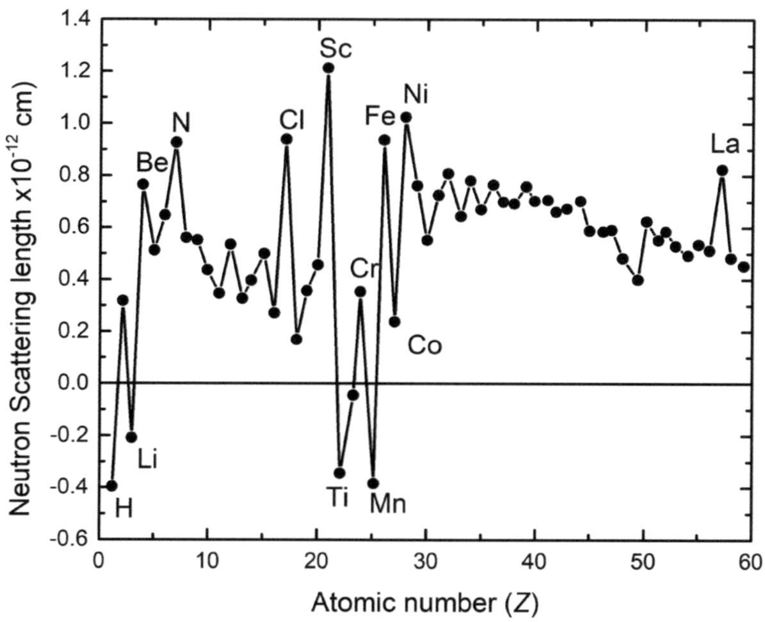

Figure 2.
The elemental scattering length behavior.

Element	Proton	Isotope	X-ray scattering ($10^{12}b_x$/cm)	Neutron scattering ($10^{12}b_{coh}$/cm)	Coherent cross section σ_{coh} (barn)	Incoherent cross section σ_{inc} (barn)	Absorption cross section σ_a (barn)
Hydrogen	1	1	0.282	−0.374	1.758	79.7, 80.27	0.33, 0.383
Hydrogen	1	2	0.282	0.667	5.592	2.0, 2.05	0.0005
Boron	5	Natural		0.530	3.54	1.70	767.0
Carbon	6	12	1.69	0.665	5.550	0.0, 0.001	0.0035, 0.004
Nitrogen	7	14	1.97	0.936, 0.940, 0.94	11.01	0.3, 0.5	1.9
Oxygen	8	16	2.16, 2.26	0.580, 0.58	4.232	0.0, 0.000	0.00019, 0.0002
Aluminum	13	Natural		0.345, 0.35	1.495	0.0, 0.008	0.23, 0.231
Silicon	14	Natural		0.42		0.0	0.17
Phosphorus	15	30	3.23	0.510			
Sulfur	16	32	4.51	0.280, 0.28			
Titanium	22	Natural		−0.344, −0.34	1.485	2.87, 3.0	6.09, 6.1
Vanadium	23	Natural		−0.038	0.018	5.07	5.08
Chromium	24	Natural		0.364	1.66	1.83	3.05
Manganese	25	55 (Natural)		−0.373	1.75	0.4	13.3
Iron	26	Natural		0.945, 0.95	11.22	0.4	2.56, 2.6
Nickel	28	Natural		1.03	13.3	5.2	4.49
Copper	29	Natural		0.772	7.485	0.55	3.78
Zirconium	40	Natural		0.716, 0.72	6.44	0.02, 0.3	0.18, 0.185
Niobium	41	93 (Natural)		0.7054	6.253	0.0024	1.15
Molybdenum	42	Natural		0.672	5.67	0.04	2.48
Cadmium	48	Natural		0.487	3.04	3.46	2520
Tin	50	Natural		0.623	4.87	0.022	0.626
Cerium	58	Natural		0.48		0.0	0.63
Gadolinium	64	Natural		0.65	29.3	151	49,700
Tantalum	73	Natural		0.691	6.00	0.01	20.6

Table 1.
Neutron scattering length and cross sections of certain elements along with their isotopes.

giving the indication is one way in which these can be distinctly identified. Other potential methods would be to study the magnetic scattering as a function of temperature or the use of polarized neutron scattering.

For certain nuclei the neutron actually interacts with the nuclide and forms a short-lived compound nucleus, which results in isotropic incoherent scattering with a negative scattering length; such interaction may potentially be associated with alteration in neutron spin. Neutron scattering by hydrogen atom falls in a special case where incoherent scattering is especially strong because the scattering proton can have either the same or opposite spin as compared to the scattered neutron [16].

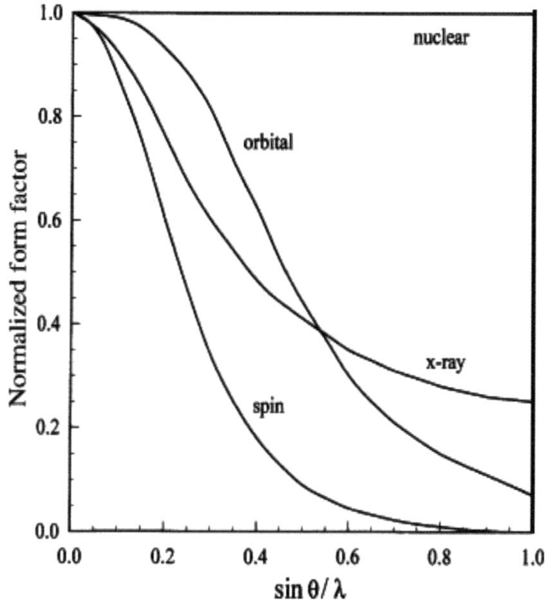

Figure 3.
The scattering length dependence on scattering angle.

These two different types, singlet or triplet scattering, lead to strong incoherent scattering by hydrogen. Interestingly, this does not apply to neutron scattering by deuterium (which is an isotope of hydrogen), and due to this reason, many neutron diffraction studies require the replacement of hydrogen by deuterium.

Another interesting aspect is pertaining to the isotope behavior; for most of the isotope atoms, the neutron scattering lengths and cross sections usually have different values as compared to their parent atom. This is in fact an added advantage which assists in understanding the compound with neutron-absorbing materials or getting the contrast in the intensity. Here we have reproduced the scattering lengths and cross section of some important atoms along with their isotopes in **Table 1**. In order to explain it, we have considered the case of nickel (Ni); the values for the different isotopes of nickel are given in **Table 1** along with that of natural abundance nickel. The differences in scattering lengths for nickel isotopes given in **Table 1** show that different but pure isotopes may be utilized to provide contrast with other elements in a compound, such contrast in the intensity is not possible to obtain with X-ray diffraction [16].

5. Neutron diffraction as probe in studying nonstoichiometry in superconductivity

The advent of cuprate-based superconductors in the late 1980s and early 1990s and the superconductivity in iron-based superconductors in the first decade of the twenty-first century created the flurry in the research in such compounds. In some of the superconducting series, the oxygen deficiencies created a significant role, and neutron diffraction played a significant role in identifying the extent of oxygen. Many superconductors are nonstoichiometric. One such example of yttrium barium copper oxide superconductor was described in earlier section.

In [17], Rupp et al. have mentioned the importance of neutron diffraction in cuprate superconductors. Their neutron diffraction data on YBCO revealed that on approaching toward low equilibrium oxygen concentrations, two different structural effects occur concurrently: first the Cu-O plane-chain bond length increases without a significant alteration in the planes or significant vacancies on the O(1) apex atom site, and second the O(4) atoms point out toward large anisotropic vibrations perpendicular to the b-axis or large static displacements from the ideal position on the b-axis. It is also pointed out that the Cu-O planes does not get affected on approaching low oxygen concentrations; it was further emphasized that in YBCO, the whole coupled Cu-O plane-chain system is responsible for the superconducting properties.

In another study by Ohshima et al., on [18] $TlSr_2CuO_y$, they could model the crystal structure of $TlSr_2CuO_y$ as orthorhombic; they also found that oxygen vacancies are formed along the *a* direction in a CuO_{2-z} sheet, which is consistent with a superstructure of doubling of the subcell along the b-direction. It was also found out that the structure involves CuO_6 octahedra and CuO4 square planes alternating along the b-direction by sharing their corners in accordance with TlO ordered oxygen vacancies in the CuO_{2-x} layer as has been presented in **Figure 4** as suggested by authors. An apical oxygen atom in the CuO6 octahedron is closer to a TlO layer than an oxygen atom shared by a TlO_6 octahedron and the CuO_4 plane as has been depicted in **Figure 5** here. The oxidation state of Cu in $TlSr_2CuO_y$ was evaluated to be 2.12 from the neutron diffraction analysis.

In [19] the simultaneous addition of CaO and CuO along with Dy3+ substitution in nonsuperconducting $La_2Ba_2Cu_4O_z$ compound started showing the signature of superconductivity. Rayaprol et al. used room temperature neutron diffraction in this resultant compound $La_{2-x}Dy_xCa_{2x}Ba_2Cu_{4+2x}O_z$ (or written as La 125 further in the text) and ascertained that the holes at La^{3+} sites are created by substitution of Ca^{2+} at La^{3+} which led to bring the two CuO_2 sheets and hence induced the superconductivity in the La-2125 compound. The typical obtained crystal structure is shown in **Figure 6**.

Tranquada et al. and reference therein in their current perspective on superconductivity, antiferromagnetism, and neutron scattering [20] have also emphasized the importance of neutron scattering in superconductors. They underlined the fact that the high-temperature superconductivity in both the copper oxide and the iron-pnictide/chalcogenide systems occurs in close proximity to antiferromagnetically ordered states. Neutron scattering has proven to be an essential technique for characterizing the spin correlations in the antiferromagnetic phases and for demonstrating how the spin fluctuations persist in the superconductors. While the nature of the spin correlations in the superconductors remains controversial, the neutron scattering measurements of magnetic excitations over broad ranges of energy and momentum transfers provide important constraints on the theoretical options.

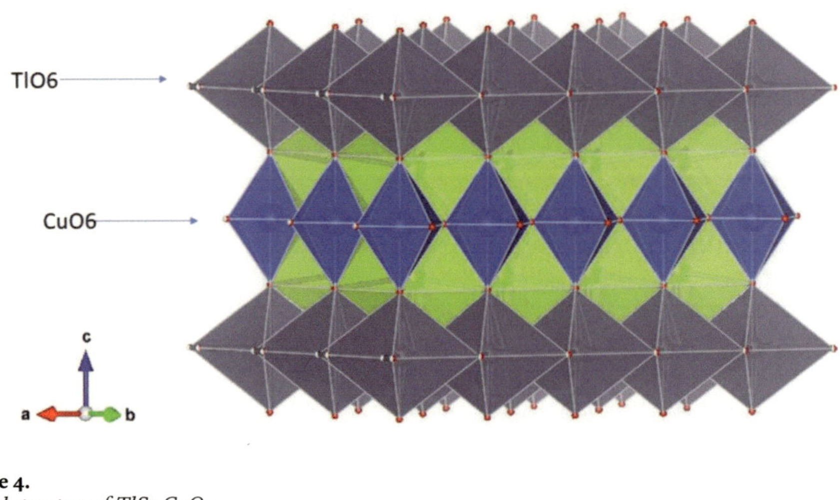

Figure 4.
Crystal structure of $TlSr_2CuO_y$.

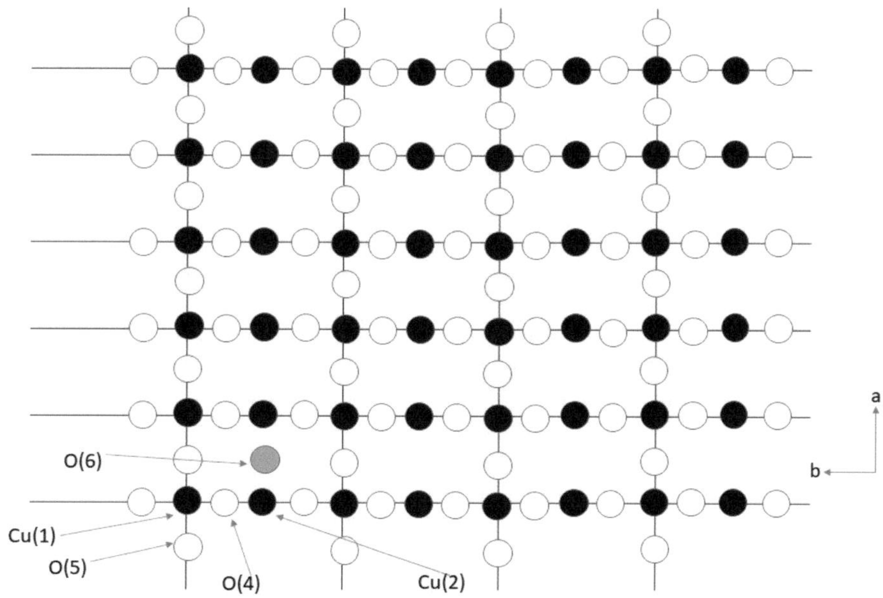

Figure 5.
Copper and oxygen arrangement on CuO_{2-x} layer in $TlSr_2CuO_y$.

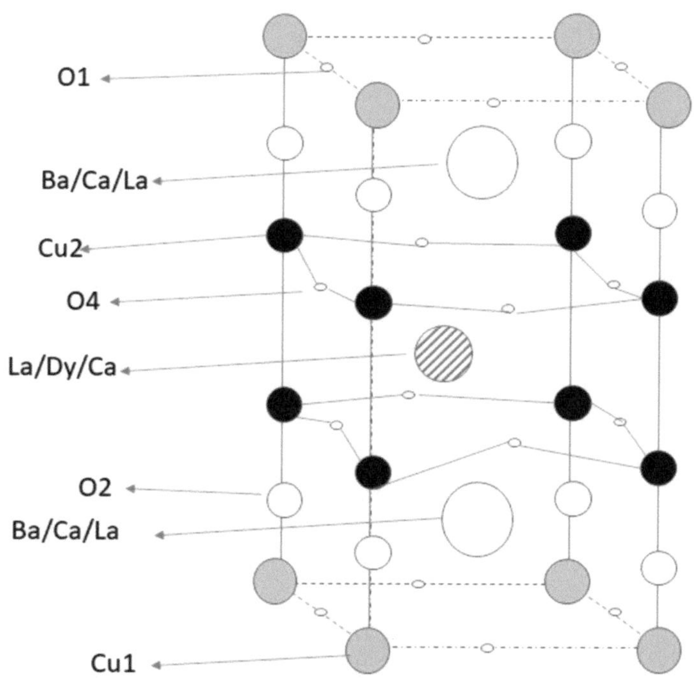

Figure 6.
La-2125 unit cell showing the CuO layer and oxygen atom position.

In a recent study on iron-based superconductor, the parent LnOFeAs (Ln = La, Sm, Ce, Nd and Gd) was found to show spin density wave below 150 K due to Fermi surface associated with tetragonal structure to transform into orthorhombic symmetry. Addition of the electron into Ln(O, F)FeAs or a hole in the (La,Sr) OFeAs layer modify the Fermi surface; thus superconductivity could be introduced.

The neutron scattering study helped in calculating the electron–phonon coupling in $LaO_{0.87}F_{0.13}FeAs$ sample; the spin density wave (SDW) could not be detected [21]. So with the help of neutron scattering experiment, it could be established that the order parameters pertaining to SDW and superconductivity are competing for itinerant holes and electrons on the Fermi surface in La(O,F)FeAs compound.

Some more examples: The structure and chemical order of a Ni-Mn-Ga Heusler alloy of nonstoichiometric composition were studied using constant-wavelength (1.538 Å) neutron diffraction at 363 K, and the diffraction pattern was analyzed. At this temperature the structure is austenite (cubic) with Fm space group and lattice constant of a = 5.83913(4) [Å]. The chemical order is of critical importance in these alloys, as Mn becomes antiferromagnetic when the atoms are closer than the radius of the 3d shell. In the studied alloy, the refinement of the site occupancy showed that the 4b (Ga site) contained as much as 22% Mn; that significantly alters the distances between the Mn atoms in the crystal and, as a result, also the exchange energy between some of the Mn atoms. Based on the analysis, the composition was determined to be $Ni_{1.91}Mn_{1.29}Ga_{0.8}$ [22].

In [23], Paulus et al. have used single-crystal neutron diffraction technique to understand the structural evolution of electrochemically oxidized $La2NiO_{4+\delta}$ crystals as a function of temperature and the oxygen intercalation and to better understand the disorder of the apical and interstitial oxygen atoms. The structural evidence were found not to attribute a peroxide character between the apical and interstitial oxygen atoms situated in exactly (1/4 1/4 1/4) of the F4/mmm unit cell. They could also reinvestigate the structural phase transition of stoichiometric La_2NiO_4 crystals at 74 K. It was also observed that the symmetry changes from Bmab to Pccn and remains orthorhombic down to 5 K.

6. Conclusion

In summary, we have explained stoichiometry and nonstoichiometry in the compounds along with some of such examples and their applications. These interesting compounds are also explained from the structural point of view. The characterization of such compound in itself is interesting science. Employing neutron to probe such compound has proven to be an effective tool in identifying this phenomenon with more clarity. In order to understand the technique, the interaction with matter was discussed in brief. The importance of these neutron-based diffraction and scattering techniques has been explained with the help of some examples discussed above.

Author details

Som Datta Kaushik*and Anil Kumar Singh
UGC-DAE Consortium for Scientific Research Mumbai Centre, BARC, Mumbai, India

*Address all correspondence to: sdkaushik@csr.res.in

References

[1] Khitarov DN, Sobolev BP, Alexeeva IV. The Rare Earth Trifluorides, Part 2 Arxius de les Seccions de Ciències. Barcelona: Institut d'Estudis Catalans, per a aquesta edicio Carrer del Carme; 2000. ISBN: 847283610X, ISBN: 9788472836105

[2] Leicester HM. The Historical Background of Chemistry. New York: Dover Publications, Inc.; 1971. ISBN: 9780486610535

[3] Geng HY, Song HX, Wu Q. Anomalies in nonstoichiometric uranium dioxide induced by a pseudo phase transition of point defects. Physical Review B. 2012;**85**(14):144111

[4] Barnes HL. Geochemistry of Hydrothermal Ore Deposits. New York: John Wiley Publications; 1997. pp. 382-390. ISBN: 978-0-471-57144-5

[5] Atkins PW, Overton TL, Rourke JP, Weller MT, Armstrong FA. Shriver and Atkins' Inorganic Chemistry. 5th ed. Great Britain: Oxford University Press; 2010

[6] Schwartz LH, Cohen JB. Diffraction from Materials. New York: Academic Press; 1977

[7] Klug HP, Alexander LE. X-Ray Diffraction Procedures. 2nd ed. New York: John Wiley; 1974

[8] Yound RA, editor. The Rietveld Method. Oxford Science Publications, Oxford University Press; 1993

[9] Dann SE. Reaction and Characterization of Solids. Cambridge: RSC; 2000

[10] Educational Material for the "Praktikum für Fortgeschrittene", Institute of Inorganic Chemistry of the University Kiel. Helmhotz-Zentrum für Materialien und Energie. Available from: https://www.uni-saarland.de

[11] Handout. Weiterbildung Chemische Nanotechnologie. Universität des Saarlands, Saarbrücken. Available from: http://www.fhi-berlin.mpg.de

[12] Handout. Application of Neutrons and Synchrotron Radiation in Engineering Material Science. Virtual Institute Photon and Neutrons for Advanced Materials (PNAM)

[13] Langford JI, Boultif A, Auffrédic JP, Louër D. The use of pattern decomposition to study the combined X-ray diffraction effects of crystallite size and stacking faults in ex-oxalate zinc oxide. Journal of Applied Crystallography. 1993;**26**:22

[14] Girgsdies F. Lecture Series Modern Methods in Heterogeneous Catalysis: X-Ray Diffraction in Catalysis. 2006. Available from: http://www.fhi-berlin.mpg.de/acnew/department/pages/teaching/pages/

[15] Mitchell PCH et al. Vibrational Spectroscopy with Neutrons, with Applications in Chemistry, Biology, Materials Science and Catalysis. London: World Scientific; 2005. 642p

[16] Sears VF. Neutron scattering lengths and cross sections. Neutron News. 1992;**3**:26. Available from: https://www.ncnr.nist.gov/resources/n-lengths/

[17] Rupp B, Fischer F, Porschke E, Arons RR, meuffels P. Neutron diffraction study of highly oxygen deficient superconductor $YBa_2Cu_3O_{6.39}$. Physica C. 1988;**156**:559

[18] Ohshima E, Kikuchi M, Izumi F, Hiraga K, Oku T, Nakajima S, et al. Structure analysis of oxygen-deficient $TlSr_2CuOy$ by neutron diffraction and

high-resolution electron microscopy. Physica C. 1994;**221**:261

[19] Rayaprol S, Kundaliya DC, Thaker CM, Kuberkar DG, Chakraborty KR, Krishna PSR, et al. Structural and magnetic studies on $La_{2-x}Dy_xCa_{2x}Ba_2Cu_{4+2x}O_z$ type superconducting oxides. Journal of Physics. Condensed Matter. 2004;**16**:6551

[20] Tranquada JM, Xu G, Zaliznyak IA. Journal of Magnetism and Magnetic Materials. 2014;**350**:148

[21] Qiu Y, Kofu M, Bao W, Lee SH, Huang Q, Yildirim T, et al. Neutron-scattering study of the oxypnictide superconductor $LaFeAsO_{0.87}Fe_{0.13}$. Physical Review B. 2008;**78**:052508

[22] Ari-Gur P et al. Neutron Diffraction Study of a Non-Stoichiometric Ni-Mn-Ga MSM Alloy. Materials Science Forum. 2013;**738-739**:103-107. Available from: https://doi.org/10.4028/www.scientific.net/MSF.738-739.103

[23] Paulus W, Cousson A, Dhalenne G, Berthon J, Revcolevschi A, Hosoya S, et al. Neutron diffraction studies of stoichiometric and oxygen intercalated La_2NiO_4. Solid State Sciences. 2002;**4**:565

Chapter 3

On Application of Hyperfree Energy for the Description of Thermodynamics of Mobile Components in Nonstoichiometric Partially Open Ceramic Systems

Jaroslav Šesták

Abstract

Nonstoichiometric oxides form a new chapter in tailored materials. Founding and construction of thermodynamic functions related to solid (geologic, metallurgic) materials is traced showing interactions between Czech Professor F. Wald and Russians R.S. Kurnakov and D.S. Korzhinskiĭ and further developed by Czech P. Holba in the initial phase definition and related characterization of partially open systems. A gradual increase in thermodynamic concepts related to solid-state description is investigated in more detail. For the associated thermodynamic definition of the mobile component, the previously formulated hyperfree energy function, which was recently applied to several systems, was used. As a measure of the material disposition for the absorption of the free component, an innovative term of plutability is proposed, which allows the introduction of various forecaster variables such as temperature, pressure, and activity. Examples of practical application are examples of high-temperature superconducting materials, where the Czech school of thermodynamics is emphasized.

Keywords: stoichiometry, ceramics, hyperfree energy, superconductors, quasimolar, plutability

1. Introduction and historical roots

While preparing a new kind of materials, the scientists often face inquiry of adjusting new thermodynamic predicaments. The recently emerging subject of oxides exhibiting variable stoichiometry is included with associated phase equilibria in relation to its environment. It influenced the new field of thermodynamics of nonstoichiometric solids, whose field was also promoted by Czech studies and whose typical cases may be represented by either spinels [1, 2], superconductors [3–6], magnetoresistance manganites [7, 8] or thermoelectric cobaltites [9–11] and all of which requiring a new approach to treating interactions between ceramic material in solid and atmospheric oxygen. It is worth of taking a notice of the core of the associated innovative thermodynamic approach, was provided by Czech physical chemists Pavel Holba [12–16], as a continuation of the recognized and previously encouraged Czech school of solid-state thermodynamics launched by Vladinir Šatava (1924–2012) [17, 18], whose worldwide international impact deserves more attention (**Figure 1**).

It was well-known that it was Josiah W. Gibbs (1876–1878), who introduced an important concept of component, chemical potential, and phase along with the ingeniously declared the phase rule [19] famous with dictum, citing: "the whole is simpler that its parts," which was genially extended in the book by Ivo Proks [20]. However, this remarkable inventiveness can be traced back to some Czech-Russian reflections relating the concepts of phases, compounds, and solutions, as it was first historically introduced by the Czech chemist František A. Wald (1861–1930). In particular, Wald's consideration [21, 22] led in the year 1912, the Russian R.S. Kurnakov (1860–1941) [23] to develop and publish a definition in which so far customary compounds were distinguished twofold as daltonides and berthollides [23–25], the latter factually being nonstoichiometric compounds. This novel idea was sustained via a more detailed examination of nonstoichiometric oxides by Yu. D. Tretyakov (1931–2012) [26]. It should be noted that Kurnakov's student and co-author of the article in which the name berthollide was first used [24] was A.I. Glazunov (1888–1951), who after World War I immigrated to former Czechoslovakia. Quaintly, he became the Rector of the Czech-Slovak Mining Academy, which was founded in 1849 in the town of Příbram.

The care for improving iron and steel products associated with various phase-composition studies of metallic samples has found a significant support in phase-diagram theory than that developed by H.E. Roozeboom (1854–1907). He was capable to apply the Gibbs phase rule in the direction of alloys (1900) [27], writing moreover a fundamental book on phase equilibria [28]. The remarkable Lewis' important concept of "activity" (1907) [29] brought a more practical form of chemical strength toward a component as well as a practical depiction of so-called: "negative common logarithm of H^+ activity" suggested, however, by Sørensen 2 years later (1909) [30] as the today's known pH quantity. In 1923, the subsequent advancement of improved thermodynamic description was accomplished by Lewis in his new book [31].

Captivatingly, Korzhinskiy nn 1936 assumed the differentiation of two classes of components useful in geochemical systems [32]. Thus, they were *inert components*

Figure 1.
The personalities staying behind the development of the so-called Czech thermodynamic school applied to the solid-state. From left: Pavel Holba (1940–2016), Vladimír Šatava (1924–2012) and Jaroslav Šesták (1938-).

whose amounts (e.g., weight or number of moles) remain independent of the amount of any other component and processes within the system, and entirely *mobile components* whose chemical potentials (and/or other intensive parameters) remain independent of equilibrium. All these concepts formed the basis for a series of studies on real material systems in a number of fields of chemistry, geochemistry, metallurgy, hydrochemistry, and material sciences. Further efforts to overcome the traditional limits on fully enclosed systems were made by Ellingham in 1944 [33]. He formed relationships between decomposition temperatures of various metal oxides in the renowned graph ΔG vs. T. In 1948, there were Richardson and Jeffes [34], which improved the traditional Ellingham diagrams by adding a so-called *monographic scale* that allowed reading the decomposition temperature of metal oxide under different types of atmospheres consisting of different gas mixtures such as O_2 + inert gas, CO + CO_2 and/or H_2 + H_2O. It looks like that the so-called "Richardson diagrams" come into sight 3 years after the foremost publication by Darken and Gurry [35, 36], dealing with the nonstoichiometry equilibrium of wüstite describing equilibria for the condensed Fe-O phases under the controlled gaseous atmosphere. Captivatingly, the experimental process by using a new gas-mixing technique was pioneering for new incoming techniques. A remarkable result in this stimulus was an important book by Muan and Osborn [37], which presented phase diagrams obtained under partially open-system conditions, for example, either in the air atmosphere or in contact with liquid iron. A distinctive application that actually surpasses the standard concepts of closed system display was established by Pourbaix in 1946 [38]. He proposed new diagrams consisting of a graph of redox potential (E) versus acid–base potential (pH) to appropriately show the equilibrium of the diluted aqueous solution at room temperatures. The coupled application has become widely known in hydrochemistry most likely due to Garrels [39] and Garrels and Christ [40]. In addition to the above-mentioned English-written books, Korzhinskiĭ has also helped to cultivate a physicochemical basis for the analysis of mineral paragenesis, which was revealed it in 1957 by a Russian book [41] also translated into English in 1959 [42].

Worth noting again is the special Korzhinskiĭ's differentiation of two classes of components in geochemical systems developed as early as in 1936 [14]. It were sorted *as inert components*, whose amounts (such as masses or numbers of moles) are independent of the amounts of any other component as well as of any processes inside the system, and *perfectly mobile components*, whose chemical potentials (or other intensive parameters) are independent factors of equilibrium [43, 44]. Afterward, he was making available the concept of thermodynamic potentials for systems containing entirely mobile components. However, requests for his new potential were first rejected by geologist Nikolaev [45], but Verhoogen [46] and Thompson [47], as well as by Palatnik and Landau [48, 49], supported Korzhinski's ideas. Later in 1956, Korzhinskiĭ published the derivation of thermodynamic potentials for systems with mobile components in the more internationally respected Russian journal [50]. It seems clear that over the past 50 years, no stimulating information on Korzhinsky's systems has been made to motivate mobile components infiltrating the geologist world into the chemistry realms. Bulach [51], Tretyakov [26, 52], and the Czech textbooks [17, 53] are weighed on Russian books, where the previous [51] contained a special chapter on "Phase diagrams for open systems." In the 1970s, the study of the equilibrium of oxides with the atmosphere also stimulated industry's interest in the production of new ceramic materials (e.g. ferrites [1, 26]) requiring properties consistent with the process of preparing new high-temperature superconductors. Accordingly, the new thermodynamic school was developed in cooperation between the Czech Academy of Sciences and the Institute of Chemical Technology in Prague, which at that time was prevented by political repression [54] but revealed in the 1990s [14–16]. Studies of oxide high-temperature superconductors [3–7] then sustained followed by other attractive ceramics, most recently the cobaltite sphere [9–11], which is characterized by its thermoelectric properties. Despite so much work that has been focused on these innovative systems that are interested in variable solid-solid gas interaction [1–11], preferably superconductor stoichiometry [5, 55–61], there is neither sufficient experimental technique nor suitable thermodynamic distinction toward to the theory of partially open systems have become accessible [15, 16], so some real cases are worth example. First, we will show some unique experimental thermoanalytical studies as a special future-oriented experimental solution.

2. Special thermoanalytical measurements applied to nonstoichiometric solids: superconductor example

Alongside ordinary low-temperature magnetic measurements [60] used to determine basic behavior of superconductivity, another specific thermoanalytical technique becomes available elucidating special features of structural behavior. One of such yet unusual diagnostic probe technique is called emanation thermal analysis (ETA) and is based on the specific measurement of radon release. Such a treatment becomes purposely beneficial in the determining fine oxygen changes [61], cf. **Figure 2**.

Beside the traditional methods based on the mass-changes detection, that is, thermogravimetry—TG or evolved gas analysis—EGA, (see left, thick curves) and their capacity to provide rather insensitive global traces, the method of emanation thermal analysis (upper) brings into play another evaluative possibility. The sample, which is regularly labeled by radioactive thorium, is acting as a recurring source for argon that diffuses and escapes finding the easiest path, so that serving as a well-detected penetrating probe. Argon go-out pathway, forced to march

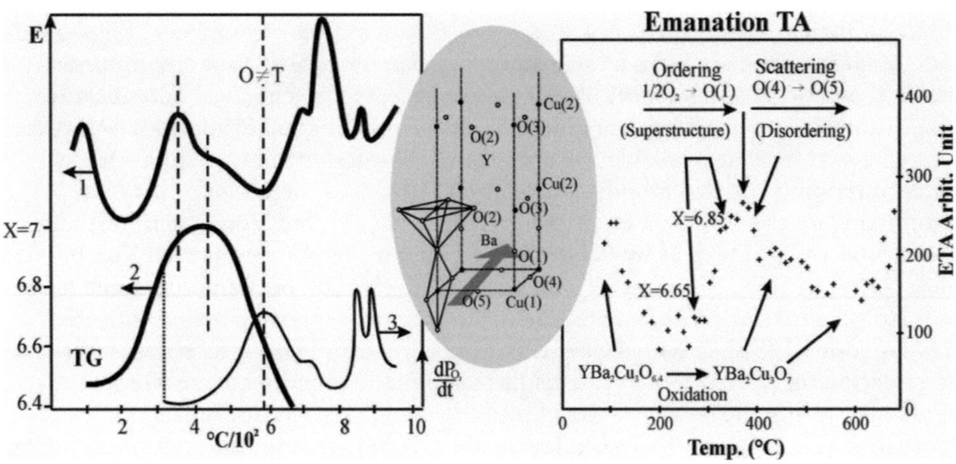

Figure 2.
Not so common portrayal for a complex analysis of $YBa_2Cu_3O_X$ behavior when exposed to oxidation changes of variable X. Beside the traditional methods of mass-changes detection (TG-2 or EGA-3, see left, thick curves bottom), the method of ETA, (upper 1) brings into play another more sensitive evaluative possibility. Equipment produced by NETZSCH ltd., (Selb, Germany) has been employed in cooperation with the Czech nuclear research Institute in Řež. The sample is labeled by radioactive thorium, which is acting as a recurring source for argon that escapes functioning as a penetrating probe (upper left and whole right). Particular ordering sequences are illustrated by arrows right.

through the sample bulk, must be receptive and big enough to accommodate such a spacious atom of argon, which is twice as big as oxygen. The best passage is certainly provided by range of defects (i.e. cracks, interfaces), but open network corridors can suit likewise. In this instance where the vacant chains are formed by O(5) voids, we can see it at the inset structure in the middle of **Figure 2**. The continual uptake of the oxygen occupation, which is enlarged along the O(4) voids, can be seen as the step-wise because the more ordered transitional structures (i.e. better superstructures) are gradually created. It reveals that the basic structure of the so-called 123-YBaCu-superconductor stays more complicated, which is closely related to the environmental conditions (temperature, pressure, or transport delays). It follows that intermediate configurations can exist between the familiar fully oxygenated (ortho-) and deoxygenated (tetra-) 123-structures, that is, the superlattice associates with δ = 6.55 [2/500] and x = 6.85 [1/300] having complete occupation every two in five and every second chain. Afterward, owing to temperature scattering, oxygen also is inclined to occupy some forbidden O(5) sites, effectively reducing thus the diffusivity free path for the probing atoms of argon through the lattice, see left diagram. Such specificity is provided by the so-called optometric measurements designed to analyze the surface/bulk heterogeneity, which helps revealing its true fractal-like dimensions [62]. For example, the distribution of pore size for $HgBa_2Ca_2Cu_3O_{8+\delta}$ was calculated from nitrogen adsorption obtaining over the size range between 1.7 and 300 nm and revealing the related fractal dimension, the value of which was calculated to lie close to 2.5.

Yet another type of experimental proof of oxygen arrangement is microscopic observation of the interfacial boundary between 123ortho- and 123tetra-transformations seen in polarized light. Diffusing oxygen builds up the orthorhombic domains with ordered Cu-O chains growing in the direction |110|, where the domain density is limited by elastic energy. Upon reaching the critical thickness, the chains begin to grow in a perpendicular direction, thereby forming a double structure as shown in **Figure 3**.

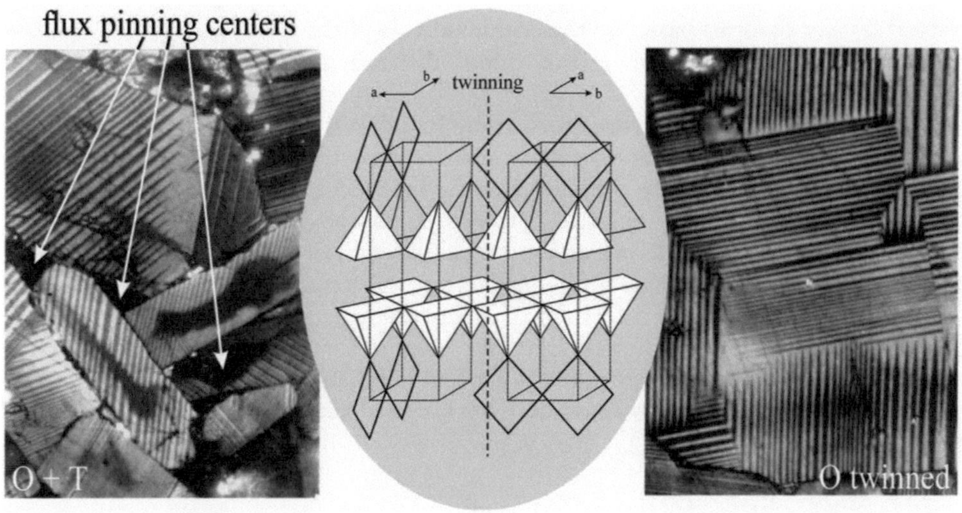

Figure 3.
Direct optical observations in polarized light are an excellent aid, showing evidence of orthorhombic domains with the ordered Cu (I)-O chains (striated). The direction of its perpendicular direction is vertically changed as shown by embedded structure, which is under the imperative impact of the energy surface providing a characteristic twinned morphology. In relation to the previous ***Figure 2****, it discloses a direct evidence of the elementary processes occurring during the oxygen uptake of $YBa_2Cu_3O_x$ and O → T transformation, which is another rather sensitive area for further research.*

Although we are often able to take care of the so-called off-stoichiometry and the incidence of a second phases in high-temperature bulk superconductors (HTSC), we also have to be aware of the peculiarities of composition at interfaces, where the high curvature $YBa_2Cu_3O_x$ crystals by depletion of oxygen (x) at the surface tension points with subsequent formation of weak bonds. Meanwhile, this can lead to a tail shift of the curve at low temperature, experimentally measured on a graph of relative conductivity versus temperature. Similar effects were previously studied due to the electric field that led to a field-controlled oxygen rearrangement [63] resulting from the movement of mobile charge carriers in the HTSC, which is therefore a generic property of these superconductors.

When conducting SQUID measurements on $YBa_2Cu_3O_x$ single crystals [60], an additional source of curiosity can be created where we can see sequential progression of a step-wise sequential dimensional transfer from the 2-D to the 3-D superconducting state. Although the first indication of superconductivity appears at a temperature of up to 92 K, internal dissipation processes hold its immediate course up to 85 K. This is because the $[CuO_2]2-$ layers are more easily rotated to superconducting state after cooling (already at T1 = 91.2 K), while by 2 K lower, it is the same with whole blocks of $\{[CuO_2]2-\ [Y]3+\ [CuO]2\}$ until a total superconductivity is achieved at even 4 K lower temperature through the Ba-CuO-Ba barrier tunnel. However, this behavior is not consistent with classical BCS theory [64], but appears to correlate more with the antiferromagnetic order in a two-dimensional system.

Whereas in the study of conventional superconductors, the main path was made possible by physical measurements, that is, discovery of superconductivity alone, Meissner effect [60, 64], isotope effect, energy gap seen first in specific heat and later in tunneling, etc. In HTSC, the most important test of new theories was performed using a large number of materials in which the CuO_2 planes are built into a completely different ionic background. Chainless materials soon eliminated the very first theories based solely on the superconductivity of CuO chains. Similarly, materials lacking the putative component eliminated apical oxygen ionization, orthorhombic distortion, phonon breeze mode, and other binding mechanisms. The critical test for theories based on the dominant role of the CuO_2 planes was the sensitivity of the critical temperature to the substitution of electronically adjacent zinc to Cu (2) positions.

Another problematic consequence can be found in the HTSC neighborhood of super-current channels, where the cationic stoichiometry may not remain stable, but its intimate bonds can be strongly modified [56, 63]. In this attitude, it can be conceptually assumed that the actual $YBa_2Cu_3O_x$ composition at $x = 6.85$ may contain both divalent and trivalent copper cations as well as holes (□), such as $YBa_2Cu_{2.6}Cu_{0.4}\square_{0.15}O_{6.85}$. Due to the negative current charge on very thin adjacent layers, it can cause not only the phase-structure deformation but also some chemical changes due to possible relocation and absorption of some species, most likely Ba^{2+} and $\square^{2+}$, which can migrate from the inner body of HTSC. The intact interface composition can then alter in an oddly wide range of stoichiometry, for example, $YBa_{3.03-3.09}Cu_{0.392-0.26}Cu_{0.4}\square_{0.223-0.226}O_{5.744-5.677}$.

3. Definition of partly open systems and the application of hyperfree energy

Lifelong experience with nonstoichiometric materials has always been in search of new expressions particularly seeking new terminology for *partly open* systems consisting of both the *free components* (proficient of exchange between the solid system and its surroundings) and *conservative components* (whose capacity is remaining constant). Conditions obligatory of the systems' isolation from its surroundings with regard to the exchanging components can be classified in analogy to the standard thermodynamic pairs: *Adiabatic–Isothermal* as well as *Isochoric–Isobaric*.

Therefore, the new-founded pair would construe as [65]: Isopletic–Isodynamic (when plethos - *πλήθος* denotes the Greek amount) and dynamis (*δύναμεις*) meaning a power, strength or ability.

Making use of the above-mentioned terms, the classification of N-component systems can be inaugurated. Two different types of components can be comprehended, that is, C—conservative (permanent) and f—free (mobile), which are satisfying thermodynamic rules C + f = N. The N-component system with f-free components and C = (N − f) conservative components can be identified as a quasi-(C)-component system, for example, the system of three components consisting of one free component, which can be called as "quasibinary system."

On continuing, we can say that it was Holba [54–56, 65–72], who implemented for partly open systems an innovative thermodynamic potential rediscovering and properly applying the correct meaning of *hyperfree energy Z*. It is defined as a Legendre transformation of the standard Gibbs free energy, *G*, with respect to a free-component amount.

$$Z = G – N_f \, \mu_f \text{ via specification } Z = f\left(T, p, \{N_c\}, \left\{\mu_f\right\}\right).$$

The above symbols are the molar content N_f of free component *f* in the system, where μ_f is its chemical potential. Because the free component *f* is understood to be shared with the surrounding atmosphere (factually serving as a reservoir), its chemical potential in the condensed system is equal to the chemical potential of surroundings. Thus, it is given as a sum of the standard chemical potential and the logarithmic term is involving activity ($\mu_f = G_f^{\circ} + RT \ln a_f$). Worth noting again is the primary Korzhinskiy's idea [44, 50] as to early distinguish between the mobile (free) components (shared with the surroundings) and the inert (conservative) components, whose pioneering thermodynamic potential μ was brought into play in the area of physical geochemistry. Unfortunately, it did not make a way into to the textbooks of solid-state chemical thermodynamics. This approach is valuable for all transformations involving interaction with the surrounding commonly studied by methods of thermal analysis [18, 73].

For a thermodynamic analysis dealing with partly open systems, it is appropriate to express the composition in terms of *quasimolar fractions* Y_i defined as $Y_i = N_i / \Sigma_c N_c = X_i / (1 - \Sigma_f X_f)$. The related hyperfree energy became thus valuable for the construction of crystal-chemical models allowing the estimation of the relations between the amount of crystal point effects and the equilibrium content of free component [65, 72] in addition to the equilibrium behavior of nonstoichiometric phases [37, 65–76]. In fact, we can observe not only closed systems where each component is isolated from the environment, but also those systems where one component (e.g. water) is exchanged with (i.e. can leak) into the environment, but other components (such as sugar or salt) remain unchanged, which become useful in our test system during processes of drying or baking.

This concept described above was further extended by deriving the related Clapeyron-like and Ehrenfest-like equations for partly open systems describing the mutual variations of predictor quantities (e.g. T-a_f, p-a_f, a_f-a_g) at the boundaries between the individual phase fields, providing an adaptable mechanism for constructing corresponding potential phase diagrams. Generally, a distinction can be made between sharp and partial phase transitions when crossing phase-field boundaries in this type of phase diagrams, which may be associated with first- and second-order phase transitions in the original Ehrenfest classification scheme, respectively. However, it should be noted that the latter should not be confused with the critical (or better lambda) phase transition described by Landau. Sharp transitions taking place in the space of variables involving the activity of free components are accompanied by abrupt changes in their content, ΔY_f. Therefore, even if the change in hyperfree energy $\Delta Z_{\varphi\text{-}\psi}$ at the boundary between phase fields φ and ψ is zero, its derivatives with respect to free-component activities $(\partial \Delta Z_{\varphi\text{-}\psi} // \partial \ln a_f)_{T,p} = \Delta Y_f$, are nonzero and we can use the same approach used to derive the standard Clapeyron equation, for example

$$\left(\frac{d \ln a_f}{d\,(1/T)}\right)_P = -\frac{(\partial(\Delta Z/T)/\partial(1/T))_{a_f,P}}{\left(\frac{\partial(\Delta Z/T)}{\partial} \ln a_f\right)_{T,P}} = \frac{\Delta H}{R\,\Delta Y_f} \tag{1}$$

$$\left(\frac{d \ln a_f}{d\ \ln a_g}\right)_{p,T} = -\frac{(\partial(\Delta Z/T)/\partial \ln a_g)_{P,T,a_f}}{(\partial(\Delta Z/T)/\partial \ln a_f)_{P,T,a_g}} = -\frac{\Delta Y_g}{\Delta Y_f} \tag{2}$$

A more intricate situation occurs when dealing with partial transitions characterized by the beginning of the precipitation of a new phase, where $\Delta Yf = 0$, since Yf is a continuous but not smooth predictor function at the boundary delimiting a partial transition, the derivatives of which are subject to sudden changes. In this line of attack, Holba called for the creation of new specialized variables [65, 72] defining a change in the composition of the free component relative to the prediction variables T, p, a_f, and a_g. Analogous to the nomenclature used for similar amounts such as isothermal compressibility or heat capacity, he proposed a new term called *plutability* (derived from the Greek πλούτος = rich, that is, susceptibility to enrichment by a given component). Based on this characterization, we can obtain variables as follows:

$$\kappa_{ff} = (\partial Y_f/\partial \ln a_f)_{T,p},\ \kappa_{fT} = (\partial Y_f/\partial \ln T)_{af,p} \text{ and } \kappa_{fp} = (\partial Y_f/\partial \ln p)_{af,T} \tag{3}$$

which can be, respectively, designated as characteristic expressions, that is *plutability*, *thermal plutability*, and *pressure plutability*.

Respecting the classical approach by Ehrenfest, we can employ the l'Hospital rule on the indefinite forms resulting from (1) and (2) and derive Ehrenfest-like equations involving the changes of plutabilities and other quantities that represent the second derivatives of the thermodynamic potential (ΔC_p, $\Delta\alpha$, $\Delta\beta$ standing for heat capacity, thermal expansion, and compressibility changes)

$$\left(\frac{d \ln a_f}{d\,(1/T)}\right)_P = \frac{\Delta C_p}{R\,\Delta\kappa_{fT}} = -\frac{TV\Delta\alpha}{R\,\Delta\kappa_{fp}} = \frac{T^2\Delta\kappa_{fT}}{\Delta\kappa_{ff}} \tag{4}$$

$$\left(\frac{d \ln a_f}{d\ \ln a_g}\right)_{p,T} = -\frac{\Delta\kappa_{gT}}{\Delta\kappa_{fT}} = -\frac{\Delta\kappa_{gp}}{\Delta\kappa_{fp}} = -\frac{\Delta\kappa_{gf}}{\Delta\kappa_{ff}} = -\frac{\Delta\kappa_{gg}}{\Delta\kappa_{fg}} \tag{5}$$

4. Practical applicability in Ca-Co-O system

Let us demonstrate the relevance of the above-derived equations on the example of Ca-Co-O system [10, 74] presented in **Figure 4**. The phase equilibria in such three-component system with a single free component (O) can be represented in a so-called quasibinary (i.e. isoactivity section) phase diagram revealed in **Figure 1b**. It exhibits exactly the same topological characteristics as conventional binary phase diagrams of closed systems. On the other hand, if we affix the cation composition and release the free-component activity, we get an isoplethal, that is, $T - \log a_f$ division presented in **Figure 4a**, for which Eqs. (1)–(4) can be employed. The upper shown two transitions, that is, ($Ca_3Co_2O_6$ + $Ca_3Co_{3.9+x}O_{9.3-\square}$) → ($Ca_3Co_2O_6$ + $Ca_{1-x}Co_xO$) and ($Ca_3Co_2O_6$ + $Ca_{1-x}Co_xO$) → $Ca_{1-x}Co_xO$ (#1 + #2), stay as the distinctive sharp transitions, where the Clapeyronian Eq. (1) is relevant. Note that if ΔY_O is constant such as in the latter case, the corresponding borderline must by linear in ($1/T$)-representation and hyperbolic in (T)-representation.

By contrast, the lower lying borderline represents a partial phase transition between the solution phase $Ca_3Co_{3.9+x}O_{9.3-\square}$ (misfit phase) and the two-phase region of this phase and $Ca_3Co_2O_6$, whose content is vanishing when crossing the boundary. Hence, the first Ehrenfestian equation must be exploited. Clearly, both ΔY_O and ΔH are zero at the transition, since the phase fraction of MF $\xi = 1$ within the single phase region and it begin to decrease at the borderline. Hence, differentiating $\Delta Y_O = (\xi - 1)\, Y_O + (1 - \xi)\,(6/5) = 0$ with respect to T and to $\ln p_O$, we obtain

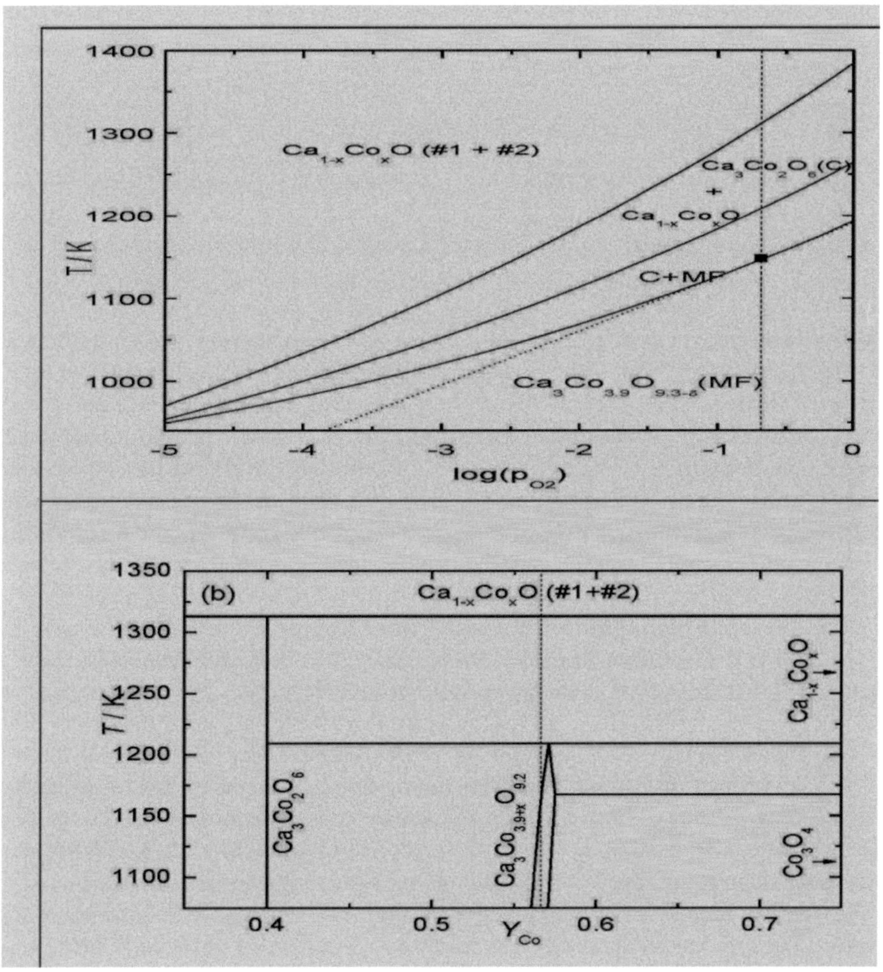

Figure 4.
(a—upper) T vs. log p_{O_2} phase diagram of Ca-Co-O system for Ca:Co = 3:3.9 (b—bottom) T versus Y_{Co} isoactivity section (p_{O_2} = 0.21) of Ca-Co-O phase diagram (dashed line—isoplethal section shown in (a)), transferred from Ref. [74].

$$\Delta\kappa_{OT} = (\xi - 1)\kappa_{OT} + \left(Y_O - \frac{6}{5}\right)\left(\frac{\partial\xi}{\partial T}\right)_{p,p_O} = \left(Y_O - \frac{6}{5}\right)\left(\frac{\partial\xi}{\partial T}\right)_{p,p_O}, \tag{6}$$

$$\Delta\kappa_{OT} = (\xi - 1)\kappa_{OO} + \left(Y_O - \frac{6}{5}\right)\left(\frac{\partial\xi}{\partial \ln p_O}\right)_{p,T} = \left(Y_O - \frac{6}{5}\right)\left(\frac{\partial\xi}{\partial \ln p_O}\right)_{p,T} \tag{7}$$

and thus

$$\frac{\Delta\kappa_{OT}}{\Delta\kappa_{OO}} = \frac{(\partial\xi/\partial T)_{p,p_O}}{(\partial\xi/\partial \ln p_O)_{p,T}} \tag{8}$$

Borderline should be linear in T provided the $\Delta\kappa_{OT}/\Delta\kappa_{OO}$ ratio was constant, which, however, is obviously not the case as seen from **Figure 4a**; the dotted line corresponds to the integrated Eq. (4) with this ratio taken constant, $\Delta\kappa_{OO}/\Delta\kappa_{OT} = 28\ K$, as obtained for p_{O_2} = 0.21 and the respective transition temperature T_t = 1148 K. Finally, the both $\Delta\kappa_{OT}$ and $\Delta\kappa_{OO}$ are accessible from experiment. Though, if $\Delta\kappa_{OT}$ can be directly evaluated from TGA data recorded at constant p_{O_2}, $\Delta\kappa_{OO}$ requires a more sophisticated device as is TGA with controlled adjustable p_{O_2} or a coulometric titration. If $\Delta\kappa_{OO}$ is not available, one can take use of another relation resulting from Eq. (4), explicitly

$$\frac{\Delta\kappa_{OT}}{\Delta\kappa_{OO}} = \frac{\Delta C_p}{RT^2\,\Delta\kappa_{OT}} = \frac{H(\mathrm{MF}) - H(\mathrm{C3C2})}{RT^2\left(Y_O - \frac{6}{5}\right)} - \frac{H_O^{\circ}}{RT^2} \tag{9}$$

having a need of the knowledge of enthalpies of both involved phases as well as gaseous oxygen at the transition temperature. This example is supposed to demonstrate how the presented equations can be applied for the construction of potential phase diagrams from experimental data or for the assessment of thermodynamic quantities necessary for the theoretical calculation of phase equilibria.

In conclusion, we can observe from the four new Clapeyronian equations (Eqs. (1) and (2) and two additional involving pressure), the providence of Holba who was able to deduce as many as 28 such pioneering Ehrenfest-like equations published in his last communications [67, 69] and finished just few days before his passing away. It is great pity that such a gifted thermodynamist could not continue publishing his concepts thus far capable of additional publications in the form of future paper.

Concisely, let us highlight that the concept of hyperfree energy and a new class of Clapeyron and Ehrenfest-like equations can be found useful for the equilibrium studies and construction of thermodynamic models of nonstoichiometric phases [65–76]. It became useful for structuring of straightforward phase diagrams reflecting the equilibrium phase relations under a given controlled atmosphere [65, 73]. The impact of Holba toward better understanding of a thermodynamic background of processes [13, 18, 65] and his encouragement of novel thermodynamic portrayal of partly open system [15, 16, 54] is unlocked to a wider attention laying emphasis on the fact that he was one of the cofounders of the renowned Czech thermodynamic school [17, 20, 72, 77, 78], cf. **Figure 1**. It is investigated in the early issues of the local Czech journal Silikaty-Ceramics published in Prague [13, 14, 18]. Holba's noteworthy legacy throw out becoming prominent within Russian circles [73, 76] and it gave stimulus for Holba's awarding the prestigious Russian Kurnakov medal of the Russian Academy of Science in the sphere of thermodynamics conferred in memoriam during the XV International Conference on Thermal Analysis held in Saint Petersburg 2016. So, this chapter is also devoted to his commemoration.

Acknowledgements

The author's assignment was exploited through the CENTEM project, reg. no. CZ.1.05/2.1.00/03.0088, co-funded by the ERDF as part of the Ministry of Education, Youth and Sports OP RDI program and, in the follow-up sustainability stage, supported through CENTEM PLUS (LO1402) by financial means from the Ministry of Education, Youth and Sports of Czech Republic carried out under the "National Sustainability Program I." The contribution of Prof. David Sedmidubský, Department of Inorganic Chemistry, University of Chemistry and Technology in Prague is highly appreciated and thoughtful attention by deceased Dr. Pavel Holba, New Technologies—Research Centre of the Westbohemian Region, University of West Bohemia in Pilsen is belatedly and copiously treasured.

Author details

Jaroslav Šesták
New Technologies - Research Centre of the Westbohemian Region, University of West Bohemia in Pilsen (NTC-ZČU), Pilsen, Czech Republic

*Address all correspondence to: sestak@fzu.cz

References

[1] Holba P, Khilla MA, Krupička S. On the miscibility gap of spinels $Mn_xFe_{3-x}O_4$. Journal of Physics and Chemistry of Solids. 1973;**34**:387-395 (later transferred to the Phase Diagrams for Ceramists, Vol. 4)

[2] Nádherný L, Jankovský O, Sofer Z, Leitner J, Martin C, Sedmidubský D. Phase equilibria in the Zn-Mn-O spinel systém. Journal of the European Ceramic Society. 2015;**35**:555-560

[3] Šesták J, Moiseev GK, Holba P, Tsagareishvili DS. Oxide-phase simulated thermodynamics of compounds auxiliary to $YBa_2Cu_3O_{7+\delta}$ superconductors. Japanese Journal of Applied Physics. 1994;**33**:97-102

[4] Nevřiva M, Kraus H, Sedmidubský D, Holba P. Phase equilibria study in the partially open Cu-(O) and Me-Cu-(O) (Me = Sr, Bi, Ba) systems. Thermochimica Acta. 1996;**282**:205-224

[5] Šesták J, Sedmidubský D, Moiseev GK. Some thermodynamic aspect of high temperature superconductor nonstoichiometry. Journal of Thermal Analysis. 1997;**48**: 1105-1122

[6] Sedmidubský M, Leitner J, Knížek K, Strejc A, Nevřiva M. Phase relations in Hg-Ba-Ca-Cu-O systém. Physica C. 2000;**341–348**(1–4):509-510

[7] Sedmidubský D, Strejc A, Nevřiva M, Leitner J, Martin C. Structural and phase relations in the Sr-Mn-O system. Solid State Phenomena. 2003;**90-91**:427-432

[8] Sedmidubský D, Nevřiva M, Leitner J, Strejc A. Single crystal growth of $Bi_{1-x}Sr_xMnO_3$—Thermodynamics and experiment. Thermochimica Acta. 2006;**450**:38-41

[9] Jankovský O, Sedmidubský D, Rubešová K, Sofer Z, Leitner J, Růžička K, et al. Structure, oxygen non-stoichiometry and thermal properties of $Bi_{1.85}Sr_2Co_{1.85}O_{7.7-\delta}$ ceramics. Thermochimica Acta. 2014;**582C**:40-45

[10] Jankovský O, Sedmidubský D, Soffe Z, Šimek P, Hejtmánek J. Thermodynamic behavior of $Ca_3Co_{3.93+x}O_{9+\delta}$ ceramics. Ceramics–Silikáty (Prague). 2012;**56**(2):139-144

[11] Jankovský O, Sedmidubský D, Sofer Z, Čapek J, Růžička K. Thermal properties and homogeneity range of $Bi_{24+x}Co_{2-x}O_{39}$ ceramics. Ceramics-Silikáty. 2013;**57**(2):83-86

[12] Holba P. On calculations of activities of chemical individuals from disorder models. Thermochimica Acta. 1972;**3**:475

[13] Holba P. Thermodynamic aspects of thermal analysis of solids. Ceramics-Silikáty (Prague). 1976;**20**:45 (in Czech)

[14] Holba P, Nevřiva M. Phase-diagram analysis of oxide systems with reset to the laws of thermodynamics. Ceramics-Silikáty (Prague). 1990;**34**:77-90

[15] Sestak J. Outline of hyperfree energy and equilibrium background opening new frontiers of thermal analysis: Inspirational links to Pavel Holba legacy. Journal of Thermal Analysis and Calorimetry. 2017;**128**:605-610

[16] Holba P. Chapter 1: Thermodynamics and ceramic systems. In: Koller A, editor. Structure and Properties of Ceramic Materials. Amsterdam/Tokyo: Elsevier; 1994. pp. 17-113

[17] Šatava V. Úvod do Fyzikální Chemie Silikátů [Introduction to the Physical Chemistry of Silicates]. Prague: SNTL; 1962 (in Czech)

[18] Holba P, Šesták J. Czechoslovak footprints in the development of

methods of calorimetry and thermal analysis. Ceramics-Silikaty (Prague). 2012;**56**:111

[19] Gibbs JW. On the equilibrium of heterogeneous substances. Transactions of the Connecticut Academy. 1876/1878; **3**:108-248/343-524

[20] Proks I. Whole Is Simpler than Its Parts: The Selected Chapters from Thermodynamic History. Bratislava: Slovak Academy of Sciences; 2012 (in Slovak)

[21] Wald F. Was ist ein chemisches Individuum [What is a chemical individual?]. Zeitschrift für Physikalische Chemie. 1899;**28**:13-16

[22] Wald F. Chemie Fasí [Chemistry of Phases]. Prague: Czech Academy of Art; 1918

[23] Kurnakov NS. Compound and chemical individual, Изв. Император. Акад. Наукъ (Bull. Acad. Imp. Sci.) St. Petersburg; 1914

[24] Kurnakov NS, Glazunov AI. Alloys of cadmium with silver and copper. Journal of the Russian Physico-chemical Society. 1912;**44**:1007

[25] Kurnakov NS. Continuousness of chemical transitions of substance. УФН. 1924;**4**(6):339-356

[26] Tretjakov JD. Thermodynamics of Ferrites. Leningrad: Izd. Chimija; 1967. 304pp. (in Russian)

[27] Bakhuis-Roozeboom HE. Eisen und stahl vom standpunkte der phasenlehre. Zeitschrift für Physiologische Chemie. 1900;**34**:437-487

[28] Bakhuis-Roozeboom HE. Die Heterogenen Gleichgewichte vom Standpunkte der Phasenlehre. Vol. I–III. Braunschweig: Fr. Vieweg & Sohn; 1901

[29] Lewis GN. Outline of a new system of thermodynamic chemistry, Proc. Am. Acad. 43 259–93 (1907). Zeitschrift für Physikalische Chemie. 1908;**61**:129

[30] Sørensen SPL. Enzym studies II. On measurement and importance of hydrogen-ion concentration at enzymatic processes. Biochemische Zeitschrift. 1909;**8**:1-153; 373-377 (in Swedish)

[31] Lewis GN, Randall M. Thermodynamics and the Free Energy of Chemical Reactions. New York: McGraw-Hill; 1923

[32] Koržinskij DS. Mobility and Inertness of Components at Metamorphosis. Vol. 1. Moscow: Izv. Akad. Nauk SSSR, Ser. Geol.; 1936

[33] Ellingham HJT. The physical chemistry of process metallurgy. Journal of the Society of Chemical Industry (London). 1944;**63**:125

[34] Richardson FD, Jeffes JHE. Free energies of formation of metal oxides as a function of temperature. The Journal of the Iron and Steel Institute. 1948;**160**: 261-273

[35] Darken LS, Gurry RW. The system iron—Oxygen. I. Journal of the Chemical Society. 1945;**67**(8):1398-1412

[36] Darken LS, Gurry RW. The system iron—Oxygen. II. Journal of the Chemical Society. 1946;**68**:798-816

[37] Muan A, Osborn EF. Phase Equilibria Among Oxides in Steelmaking. Reading, MA: Addison-Wesley; 1965. 236pp

[38] Pourbaix MJN. Thermodynamics of Dilute Aqueous Solutions. London: E. Arnold and Co.; 1949. 136pp (French original: Thermodynamique des Solutions Aqueses Diluées; 1946)

[39] Garrels RM. Mineral Equilibria. New York: Harper and Row; 1965

[40] Garrels RM, Christ CL. Solutions, Minerals, and Equilibria. New York: Harper and Row; 1965. 450p

[41] Korzhinskiĭ DS. Physicochemical Basis of the Analysis of the Paragenesis of Minerals. Moskva: Izd. Akad. nauk SSSR; 1957. (in Russian)

[42] Korzhinskiĭ DS. Physicochemical Basis of the Analysis of the Paragenesis of Minerals (English Translation). New York: Consultants Bureau; 1959

[43] Korzhinskiĭ DS. Sootnošenie meždu mineralogičeskim sostavom i veličinoj chimičeskich potencialov komponěntov [Relations between mineralogical composition and values of chemical potential of components]. Zapiski Vserossiiskogo Mineralogicheskogo Obshchestva. 1944; **73**(1):111

[44] Korzhinskiĭ DS. Termodinamičeskije Potencialy Otkrytych Sistem i Primer ich Primenenija v Geochimii [Thermodynamic Potentials of Open Systems and Example of their Application in Geochemistry]. Moscow: Izv. Sektora Fiz.-Chim. Analiza Akad. Nauk SSSR; 1949. p. 19

[45] Nikolajev VA. Ob odnom iz osnovnych uravnenij termodinamiky i ego priměněnii k otkrytym sistemam [On one of the fundamental equations of thermodynamics and on its application to open systems]. Zapiski Vserossiiskogo Mineralogicheskogo Obshchestva. 1954; **83**(4):777

[46] Verhoogen J. Thermodynamics of a magmatic gas phase. Bulletin of the Department of Geological Sciences. 1949;**28**(5):143

[47] Thompson JB. The thermodynamic basis for the mineral fancies concept. American Journal of Science. 1955; **253**(2):65-103

[48] Palatnik LS, Landau AI. K teorii ravnovesia geterogennych mnogokomponentnych system [On theory of equilibrium of heterogeneous multicomponent systems]. Doklady Akademii Nauk SSSR. 1955;**102**(1):125

[49] Palatnik LS, Landau AI. Fazovyje Ravnovesia Mnogokomponentnych Sistemach [Phase Equilibria of Multicomponent Systems]. Charkov: Izd. Charkov. Univ.; 1961. 406pp

[50] Korzhinskiĭ DS. Vyvod termodinamičeskich potencialov sistem s podvižnymi komponentami [Derivation of thermodynamic potentials of systems with Mobile components]. Doklady Akademii Nauk SSSR. 1956;**106**(2):441

[51] Bulach AG. Metody Termodinamiky v Mineralogii [Methods of Thermodynamics in Mineralogy]. Leningrad: Izd Nědra; 1968. 175pp

[52] Tretyakov YD. Химия нестехиометрических окислов [Chemistry of Nonstoichiometric Oxides]. Moskva: Izd. Mosk. Univ.; 1974

[53] Rieder M, Povondra P. Fázové Rovnováhy pro Geology [Phase Equilibria for Geologists]; Karolinum, Praha (1997). 1st ed: SPN Praha (1979). 252+66pp

[54] Holba P. Equilibria of Mn-Fe-O Spinels with gaseous oxygen and their interpretation. Czechoslovak Journal of Physics (Prague). 1992;**42**:549-575. Dissertation with 110 pp was forbidden to defending as well as publishing due to political reasons, Prague 1970 (in Czech), latter published as thermodynamics of partially open systems

[55] Šesták J. Binary and ternary compounds, phase diagrams and contaminations in the Y-Ba-Cu-O system auxialiary to superconducting

ceramics. Thermochimica Acta. 1989; **148**:235

[56] Šesták J, Kamarád J, Holba P, Tříska A, Pollert E, Nevřiva M. Charge distribution, pressure and compositional effects of CuO based superconductors. Thermochimica Acta. 1991;**174**:99

[57] Moiseev GK, Vatolin NA, Šesták J, Tsagareishvili DS. Evaluation of thermodynamic data in Y-Ba-Cu-O systems. Thermochimica Acta. 1992; **198**:267

[58] Šesták J. Phase diagram, glass formation and crystallization in the Bi-Ca-Sr-Cu-O superconductive system. Journal of Thermal Analysis. 1991;**36**: 1639

[59] Šesták J. Phase diagram of CuO based high temperature superconductors. Pure and Applied Chemistry. 1992;**64**:125

[60] Šesták J, Lipavský P. Chronicle of high-temperature oxide superconductors. Journal of Thermal Analysis and Calorimetry. 2003;**74**:365

[61] Balek V, Šesták J. Emanation thermal analysis of YBCO-superconductor preparation and characterization. Thermochimica Acta. 1988;**133**:23

[62] Staszcuk P, Sternik D, Chadzynski GW. Determination of total heterogeneity and fractal dimensions of high-temperature superconductors. Journal of Thermal Analysis and Calorimetry. 2003;**71**:173

[63] Moiseev G, Šesták J, Štěpánek B. Possible compositional changes of the $YBa_2Cu_3O_x$ surface layer on the boundary with negative charge current. Ceramics-Silikáty (Prague). 1994;**38**:143

[64] Bardeen J, Cooper LN, Schieffer JR. Theory of superconductivity. Physics Review. 1957;**108**:1175

[65] Holba P. Chapter 4: New thermodynamic potentials and Clapeyron equations for condensed partly open systems applicable to nonstoichiometric phases. In: Šesták J, Holeček M, Málek J, editors. Thermodynamic Structural and Behavioral Aspects of Materials Accentuating Non-Crystalline States. Liberec: OPS Plzeň; 2011. pp. 98-115

[66] Holba P, Sedmidubský D. Heat capacity equations for nonstoichiometric solids. Journal of Thermal Analysis and Calorimetry. 2013;**113**:239-245

[67] Holba P. Chapter 2: Equilibrium background of processes initiated by heating and Ehrenfests´ classification of phase transitions. In: Šesták J, Šimon P, editors. Thermal Analysis of Micro-Nano- and Non-crystalline Materials: Transformation, Crystallization, Kinetics and Thermodynamics. Berlin: Springer; 2013. pp. 29-52

[68] Sedmidubský D, Holba P. Material properties of nonstoichiometric solids. Journal of Thermal Analysis and Calorimetry. 2015;**1120**:183-188

[69] Holba P. Ehrenfest equations for calorimetry and dilatometry. Journal of Thermal Analysis and Calorimetry. 2015;**120**:175-181

[70] Holba P, Sedmidubský D. Chapter 3: Crystal defects and nonstoichiometry contributions to heat capacity of solids. In: Šesták J, Šimon P, editors. Thermal Analysis of Micro- Nano- and Non-crystalline Materials: Transformation, Crystallization, Kinetics and Thermodynamics. Berlin: Springer; 2013. pp. 53-74

[71] Holba P. Thermodynamic description of heat capacities in nonstoichiometric phases. Chemicke Listy. 2015;**109**:113-116 (in Czech)

[72] Holba P, Sedmidubský D. Chapter 7: Clapeyron and Ehrenfest equations and

hyperfree energy for partly open systems. In: Šesták J, Hubík P, Mareš JJ, editors. Thermal Physics and Thermal Analysis: From Macro to Micro Highlighting Thermodynamics, Kinetics and Nanomaterials. Berlin: Springer; 2017. ISBN: 978-3-319-45897-7

[73] Šesták J, Holba P, Gavrichev KS. Reinstatement of thermal analysis tradition in Russia and related East European interactions. Journal of Thermal Analysis and Calorimetry. 2015;**119**:779-784

[74] Sedmidubský D, Šesták J. On the history and recent application of hyperfree energy when describing thermodynamics of mobile components in partly open ceramic systems. Ceramics-Silikáty. 2017;**61**:250-256

[75] Šesták J. On chronicle and recent exploit of a special clone of free energy (hyper) beneficial to thermodynamic description of partially open geoleogical, metallurgical and other nonstoichiometric systems. Journal of Mining and Metallurgy, Section B: Metallurgy. 2017;**53**:249-254

[76] Šesták J. The Holba and Korzhinski treatments of hyperfree energy effective to thermodynamic of partially open solid-state systems of nonstoichiometric oxides. Glass Physics and Chemistry. 2017;**43**:283-286

[77] Šesták J. Science of Heat and Thermophysical Studies: A Generalized Approach to Thermal Analysis Thermodynamics. Amsterdam: Elsevier; 2005

[78] Šesták J. Ignoring heat inertia impairs accuracy of thermodynamic determination in thermal analysis experiments. International Journal of Chemical Kinetics. 2019;**51**:74-80

Chapter 4

Nonstoichiometry Role on the Properties of Quantum-Paraelectric Ceramics

Alexander Tkach and Paula M. Vilarinho

Abstract

Among the lead-free perovskite-structure materials, strontium titanate ($SrTiO_3$—ST) and potassium tantalate ($KTaO_3$—KT), pure or modified, are of particular importance. They are both quantum paraelectrics with high dielectric permittivity and low losses that can find application in tunable microwave devices due to a dependence of the permittivity on the electric field. Factors as Sr/Ti and K/Ta ratio in ST and KT ceramics, respectively, can alter the defect chemistry of these materials and affect the microstructure. Therefore, if properly understood, cation stoichiometry variation may be intentionally used to tailor the electrical response of electroceramics. The scientific and technological importance of the stoichiometry variation in ST and KT ceramics is reviewed and compared in this chapter. The differences in crystallographic phase assemblage, grain size, and dielectric properties are described in detail. Although sharing crystal chemical similarities, the effect of the stoichiometry is markedly different. Even if the variation of Sr/Ti and K/Ta ratios did not change the quantum-paraelectric nature of ST and KT, Sr excess impedes the grain growth and decreases the dielectric permittivity in ST ceramics, while K excess promotes the grain growth and increases the dielectric permittivity in KT ceramics.

Keywords: nonstoichiometry, perovskite, electroceramics, ferroelectrics, crystallographic phase assemblage, grain growth, dielectric spectroscopy

1. Introduction

Considering functional oxides, ferroelectrics are essential materials, being used in a wide range of applications [1, 2]. Ferroelectrics are nonlinear dielectric materials and their main characteristic is a spontaneous electric polarisation that can exist without an external electric field and can be reversed by the application of the field [1, 3]. Ferroelectricity is a temperature-dependent property, inherent to materials with a noncentrosymmetric crystal structure that is lost above the characteristic temperature designated as Curie temperature (T_0), when the material becomes centrosymmetric and paraelectric. Although ferroelectric materials hold their main functionality in the polar state (below T_0), they are also useful above T_0, in the paraelectric state [4, 5]. In this nonpolar, phase ferroelectrics are normally characterised by a real part of the dielectric permittivity ε' (or susceptibility $\chi' = \varepsilon' - 1$), which depends on temperature, according to the Curie-Weiss law (Eq. (1)):

$$\varepsilon' \approx \chi' = C/(T - T_0), \tag{1}$$

where C is the Curie constant and T_0 is the Curie temperature, above which the ferroelectric material is in the paraelectric state [1, 3]. For a second-order phase transition, the transition temperature $T_C = T_0$, but for a first-order transition $T_C > T_0$ (see **Figure 1**). As also schematically shown in **Figure 1**, the second-order transition is characterised by a smooth increase of the spontaneous polarisation P_s as the temperature decreases starting from T_C, while for the first-order transition P_s jumps for some value at T_C with a further slight increase as temperature decreases [3].

Besides the temperature, ε' depends on the applied electric field, particularly, near the phase transition. There are two parameters used for the characterisation of the dependence of the dielectric permittivity on the applied DC bias electric field: (1) tunability n, defined as the ratio of the dielectric permittivity of the material at zero electric field to that at some nonzero electric field and (2) relative tunability n_r, defined as (Eq. (2)):

$$n_r(E) = [\varepsilon'(0) - \varepsilon'(E)]/\varepsilon'(0) = (n - 1)/n, \tag{2}$$

where $\varepsilon'(0)$ is the dielectric permittivity at zero field and $\varepsilon'(E)$ is the dielectric permittivity under the applied field E [4]. Thus, permittivity is an important parameter in defining capacitors with high capacity to store electrical energy, as well as high-performance tunable microwave devices, for example, phase shifters, as components in electronically scanned phased-array antennas for communications and radar applications because of their low dielectric losses and reasonable

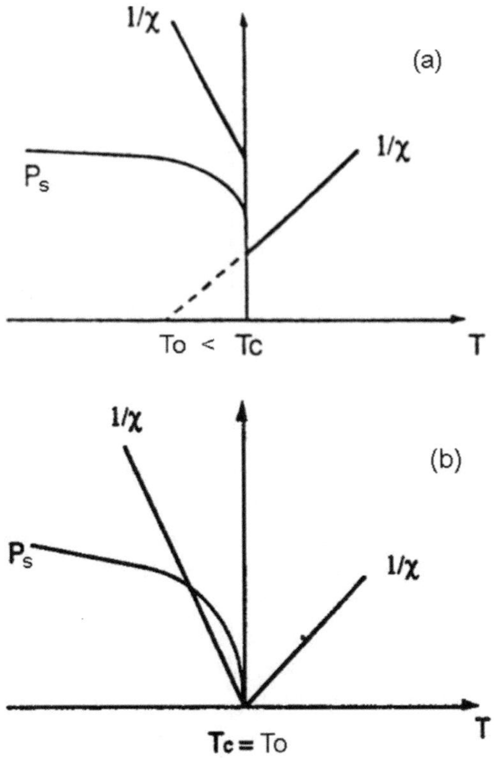

Figure 1.
Temperature dependence of inverse dielectric susceptibility χ′ and spontaneous polarisation P_s for first-order (a) and second-order (b) ferroelectric phase transitions (adapted from Smolenskii [3]).

tunability [5]. Moreover, the dielectric permittivity is a complex parameter, consisting of a real part ε' and an imaginary part ε'', while their ratio determines the dissipation factor $\tan\delta = \varepsilon''/\varepsilon'$. In this respect, incipient ferroelectrics (or quantum paraelectrics), such as perovskite type $SrTiO_3$ (ST) and $KTaO_3$ (KT), are of great interest because they do not possess the phase transition into polar phase at any temperature, simultaneously presenting very low dielectric losses. Indeed, they can possess a dissipation factor $\tan\delta$ as low as 10^{-4}–10^{-5} that is very attractive for microwave applications [6, 7].

Strontium titanate ($SrTiO_3$, ST) and potassium tantalate ($KTaO_3$, KT) belong to the family of incipient ferroelectrics because their dielectric permittivity monotonously increases upon cooling down to near 0 K without any ferroelectric-type anomaly [8]. However, since the ferroelectric order in these two materials is suppressed by quantum fluctuations, they can also be called quantum paraelectrics [9, 10], while their $\varepsilon'(T)$ dependence can be described by Barrett's relation (Eq. (3)):

$$\varepsilon'(T) = \frac{C}{\frac{T_1}{2}\coth\frac{T_1}{2T} - T_0} + \varepsilon_1 \tag{3}$$

which is based on the mean-field theory taking quantum fluctuations into account [11]. Comparing to Eq. (1) for the Curie-Weiss law, a temperature of the crossover between classical and quantum behaviour T_1 and a temperature-independent component of permittivity ε_1 are introduced in Eq. (3) for the Barrett's relation. However, in the limit $T \gg T_1$, the Barrett's relation transforms into the traditional Curie-Weiss law. The low-temperature paraelectric phase is thus unstable in quantum paraelectrics, and the ferroelectric state can be induced by the application of a high electric field [12], uniaxial stress [13], cation [10, 14–16], or oxygen isotope substitutions in the lattice [17].

Structurally, ST and KT are similar and both crystalize with a perovskite-type structure [16, 18]. The general chemical formula for the perovskite oxides is ABO_3, where A and B are cations of very different sizes (A are larger than B), and O is an oxygen that bonds to both. As shown in **Figure 2**, the perovskite unit cell is ideally cubic, where A-cations are placed at the cube corners, B-cations are located at the body centre, and the position of oxygen ions is at the centre of the faces. The cell packaging is characterised by the Goldschmidt tolerance factor (Eq. (4)):

$$t = \frac{r_A + r_O}{\sqrt{2}(r_B + r_O)} \tag{4}$$

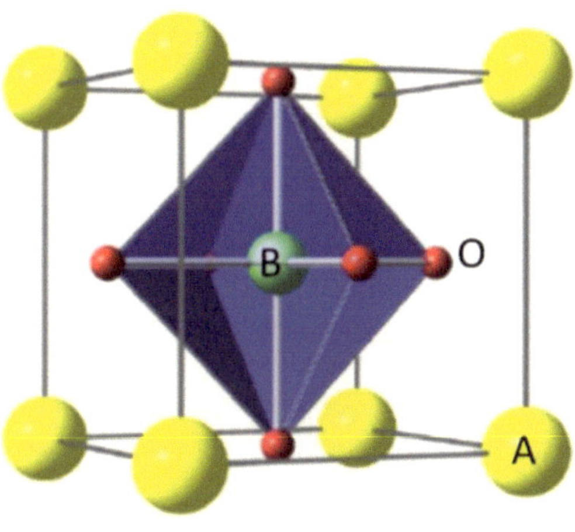

Figure 2.
Representation of the ideal cubic perovskite unit cell (ABO_3).

where r_i (i = A, B, O) denotes the average ionic radii of the constituents of ABO_3 compound in the respective sites [18, 19]. In the case of ST and KT, t is close to 1, implying that both A and B ions are closely packed and their excess solubility is very limited.

Regarding the cation excess solubility limits for the ST lattice, a presence of TiO_2 second phase was reported for ST ceramics with Ti excess down to 0.5 mol%, in agreement with similar high-temperature conductivity behaviour observed for Sr/Ti ratio ≥ 0.995 [20]. On the other hand, Sr excess is known to accommodate in the ST lattice as a three-dimensional mosaic of single-layered rock-salt blocks, forming the so-called Ruddlesden-Popper structures with the formula $SrO{\cdot}(SrTiO_3)_n$ instead of secondary phases [21]. Concerning the electrical properties, a breakdown strength was reported to be higher for ST ceramics with Sr/Ti ratio of 0.996, comparing to that for stoichiometric ones, and attributed to smaller grain size [22]. More recently, we have also investigated the effect of nonstoichiometry—Sr/Ti ratio from 0.995 to 1.02—on the high-temperature electrical response of ST ceramics, using impedance spectroscopy [23]. The resistivity of bulk and grain boundaries systematically decreased in both Ti-rich and Sr-rich ST, as compared to stoichiometric ceramics. The nonstoichiometry effect was found to be much stronger for the grain boundaries as compared to the bulk and attributed to the defect chemistry variation rather than to the microstructural development [23].

In the case of KT, in which the dielectric losses can be even lower than those of ST, thus exhibiting a dissipation factor tanδ of ~10^{-4} in the GHz range [6, 7], stoichiometry effect is even more important and, additionally, more difficult to control due to the high volatility of the alkali element as potassium [24, 25]. As a result, the dielectric properties have been mainly reported for KT single crystals [6, 7, 10, 24, 26–29], whereas the studies on polycrystalline bulk are seldom reported [24, 25, 30], even though ceramics are simpler and less expensive to produce than single crystals. This scarcity is enhanced by the fact that although KT melts easily above 1350°C, it is hard to obtain a highly dense monophasic stoichiometric polycrystalline KT below this temperature [25]. In addition, according to our pioneer thermodynamic studies using oxide melt solution calorimetry, the enthalpy of formation of perovskite from oxides becomes less exothermic from pyrochlore phases, thus indicating a less stable structure with respect to the constituent oxides. The decomposition enthalpy of $K_2Ta_2O_6$ to $KTaO_3$ indicates that pyrochlore is energetically more stable than perovskite, and also confirms that pyrochlore is the low-temperature phase [31].

Needless to state that the optimisation of the dielectric response of functional materials is evidently associated with the precise control of the composition (namely the stoichiometry). Therefore, this chapter is aimed to overview and to compare the effect of cationic ratio on the microstructural and dielectric properties of ST and KT ceramics.

2. Experimental

2.1 Preparation of ST ceramics

Ceramics of strontium titanate were prepared by conventional mixed oxide method [32]. Reagent grades $SrCO_3$ and TiO_2 were weighed according to the compositions $Sr_{1.02}TiO_{3.02}$, $SrTiO_3$ and $Sr_{0.997}TiO_{2.997}$. After milling in alcohol for 8 h using Teflon pots and zirconia balls in a planetary mill, the powders were dried, and then calcined at 1150°C for 2 h. The calcined powders were ball milled under similar conditions as the previous ones and dried again to obtain powders with particle size lower than 5 μm. Pellets of 10 mm in diameter were uniaxially pressed at 100 MPa and then isostatically pressed at 200 MPa. Sintering was performed in air at 1500°C for 5 h with heating and cooling rates of 5°C/min. The density of all the sintered samples, reached ~97% of the theoretical density of ST.

2.2 Preparation of KT ceramics

Ceramics of potassium tantalate were also prepared by the conventional mixed oxide method [33]. After being dried for dehydration, K_2CO_3 and Ta_2O_5 reagents were weighed according to the compositions $KTaO_3$, $K_{1.02}TaO_{3.01}$, and $K_{1.05}TiO_{3.025}$. Once milled in a planetary mill for 5 h using Teflon pots, zirconia balls, and alcohol, the powders were dried, and then calcined at 875°C for 8 h. The calcined powders were ball milled in alcohol for 5 h and dried again. Pellets of 10 mm in diameter were uniaxially pressed at 100 MPa, covered by powder of the same composition to decrease the loss of potassium, and sintered in closed alumina crucibles at 1350°C for 1 h with heating and cooling rates of 5°C/min. The density of all the sintered samples varied from ~87 to 90% of the theoretical density of KT. Through weight loss and inductively coupled plasma spectroscopy analysis, the potassium loss was about 3–4%.

2.3 Characterisation of the ceramics

Room temperature X-ray diffraction (XRD) analysis (Rigaku D/Max-B, Cu Kα) was conducted on some of the grounded sintered pellets with a scanning speed of 1 °/min and a step of 0.02°. Lattice parameters were refined by the least-square fitting to the observed XRD data, between 2Θ = 20° and 110°, using WinPLOTR software. The microstructure of the ceramics was observed on polished and thermally etched sections using scanning electron microscopy (SEM, Hitachi S-4100 and Hitachi SU-70). The average grain size of the sintered pellets was measured on at least 100 grains by AnalySIS (Soft Imaging System GmbH) software. For the dielectric measurements, gold electrodes were sputtered on both sides of the polished ceramics. The dielectric permittivity and loss were measured at different frequencies between 100 Hz and 1 MHz, using Precision LCR Meter HP 4284A and a Displex APD-Cryogenics cryostat of He closed cycle during heating in the temperature range from 10 to 300 K.

3. Results

3.1 Structure and microstructure

XRD patterns of the sintered ST ceramics with initial Sr/Ti ratio = 0.997, 1, and 1.02 are shown in **Figure 3** (left). From the XRD analysis, all ST compositions under study have a cubic perovskite structure and are monophasic. No systematic variation of the lattice parameter was observed.

For the sintered KT ceramics with initial K/Ta ratio = 1, 1.02, and 1.05, the XRD patterns are shown in **Figure 3** (right). The observed X-ray diffraction lines are consistent with the cubic perovskite symmetry of stoichiometric KT for all the precursor compositions. For ceramics with K/Ta = 1.05 and 1.02, no distinct secondary phases are detected. Conversely, additional diffraction lines observed in the patterns for K/Ta = 1, evidence the existence of a secondary phase, which was assigned to the potassium-poor tungsten bronze structure $K_6Ta_{10.8}O_{30}$ phase. These results are in agreement with an homogeneous distribution of both potassium and tantalum

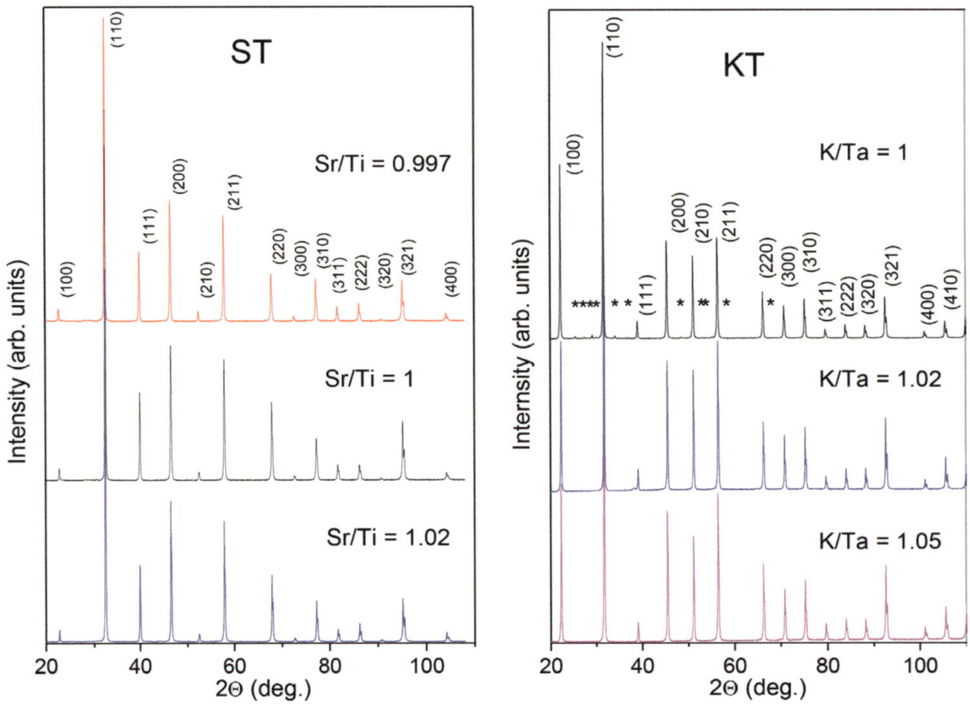

Figure 3.
*XRD patterns of sintered strontium titanate ($SrTiO_3$, ST) (left) and potassium tantalate and potassium tantalate ($KTaO_3$, KT) (right) ceramics, prepared with indicated Sr/Ti and K/Ta ratios (adapted from [32, 33]). Reflections of ST and KT phases are marked by their corresponding crystallographic indexes and reflections of $K_6Ta_{10.8}O_{30}$ secondary phase are marked by *.*

in the grains of KT ceramics with initial K/Ta ratio of 1.05 and 1.02, observed by elemental mapping using energy dispersive spectroscopy, while some Ta-rich areas were detectable in the ceramics with the initial K/Ta = 1 [33]. Moreover, since no secondary phase was detected in the XRD patterns of KT powders after calcination (not shown) for all the precursor compositions including K/Ta = 1, it is assumed that the sintering process at 1350°C mainly leads to the loss of volatile potassium. The lattice parameter values of KT phase deduced from the XRD patterns were close to that of 3.989 Å for KT single crystals [26].

Rather dense microstructures and significant difference in the grain size for Ti-rich and Sr-rich ST ceramics was observed by scanning electron microscopy [32]. The average grain size of ST ceramics with Sr/Ti ≤ 1 was found to be of about 20 μm, that is, in the range of tens of microns, whereas Sr/Ti > 1 yields ceramics with the grain size of about 6 μm, that is, in the micron range (see **Table 1**). The microstructural analysis of KT ceramics revealed cubic-like grain shape and well-defined porosity [32] in agreement with the ceramics relative density of about 88%. Moreover, the grain size was found to grow from submicron to several microns range with K/Ta ratio increasing from 1 to >1. Average grain-size values of 0.7, 4.9, and 6.5 μm were determined for the ceramics with initial K/Ta ratio of 1, 1.02, and 1.05, respectively, as also displayed in **Table 1**.

Thus, the grain-size dependence on the stoichiometry of ST and KT ceramics behaves oppositely. The larger grains are formed for excess of B-site cations in ST and for excess of A-site cations in KT. Such dissimilarity is based on the unique crystallochemistry details of each system, as displayed by their phase diagram. An eutectic liquid phase that promotes the grain growth during the sintering exists on the Ti-rich side of the $SrO\text{-}TiO_2$ phase diagram, when Sr/Ti < 1 [34]. In contrast,

Ceramics	A/B ratio	Average grain size, μm	ε'_{max}	$\tan\delta_{min}$, %	Barrett relation parameters			
					T_0, K	T_1, K	$C/10^3$, K	ε_1
ST	0.997	20	~7700	0.40	34	98	112	—
	1	20	~6300	0.33	35	99	92	—
	1.02	6.0	~3900	0.69	32	110	87	—
KT	1	0.7	~2250	0.34	14	66	38	123
	1.02	4.9	~4000	0.25	12	48	49	58
	1.05	6.5	~4000	0.62	10	48	57	120

Table 1.
Average grain size and Barrett's relation parameters for $SrTiO_3$, ST ceramics with initial Sr/Ti ratio of 1, 1.02, and 0.997 sintered at 1500°C for 5 h and for $KTaO_3$, KT ceramics with initial K/Ta ratio = 1, 1.02, and 1.05 sintered at 1350°C for 1 h [32, 33].

the KT grain boundaries become wet close to the eutectic temperature that emerges for K/Ta > 1 [35]. Then, grain boundary diffusion increases and grain growth is promoted in the presence of potassium excess.

3.2 Electrical properties

The low-frequency dielectric measurements data are summarised in **Figures 4** and **5**. The temperature dependence of the dielectric permittivity (ε') for stoichiometric and nonstoichiometric ST ceramics, which revealed no considerable frequency dispersion is shown in **Figure 4a** for the frequency of 10 kHz. No dielectric permittivity anomaly was observed, as well. The dielectric permittivity increases steeply and levels-off at high values as the temperature approaches 0 K, revealing a typical behaviour of quantum paraelectrics [9, 10]. Comparing with stoichiometric composition, Sr excess lowers the dielectric permittivity values at low temperatures from ~6300 to ~3900, while Ti excess raises it to ~7700, as also listed in **Table 1**. The later value is lower than that of ~20,000, reported for ST single crystals [9], due to the contribution of pores and grain boundaries with much lower permittivity than that of ST bulk [36]. However, it is much higher than that of ~5600 for conventionally prepared ST ceramics sintered at lower temperature of 1400°C [37] and even higher than that of ~6850, reported for higher purity sol-gel derived ST ceramics [38].

The temperature dependence of the dielectric permittivity of KT ceramics, with initial K/Ta ratio = 1, 1.02, and 1.05 at the frequency of 10 kHz is shown in **Figure 4b**, revealing too the continuous increase of $\varepsilon'(T)$ on cooling. The potassium excess is found to raise the dielectric permittivity monotonously in the range of initial K/Ta ratio under study, in spite of the small difference between the low-temperature dielectric permittivity for K/Ta = 1.02 and K/Ta = 1.05. For KT ceramics with initial K/Ta ratio of 1, the dielectric permittivity is about 2300. It is much smaller than that of KT single crystals [7] but close to that reported for KT ceramics [24]. Contributions of secondary phases in the ceramics, which are less polarisable than crystalline grains of KT, grain boundaries (small grain size), and pores (density ~88%) can lead to the observed permittivity decrease. On the other hand, the dielectric permittivity of similarly dense KT ceramics with initial K/Ta ratio of 1.02 and 1.05 reach a value of about 4000, which is much higher than the values, reported for K/Ta = 1 by Chen et al. [24], higher than the value of 3100, reported for K/Ta = 1.05 by Axelsson et al. [25], and is not much below 5000,

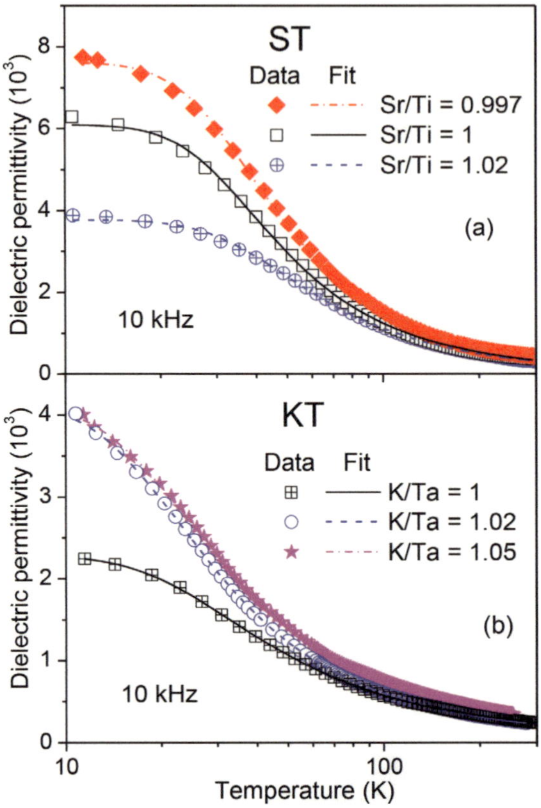

Figure 4.
Temperature dependence of the real part of the dielectric permittivity ε′ of strontium titanate ($SrTiO_3$, ST) (a) and potassium tantalate ($KTaO_3$, KT) (b) ceramics with indicated Sr/Ti and K/Ta ratios at frequency of 10 kHz (adapted from [32, 33]). The fit curves of the experimental data to the Barrett's relation are shown as well.

which was reported for KT single crystals [7]. In parallel to our work, Glinšek et al., obtained monophasic KT ceramics with relative density ≥95% using hot-pressing of mechanically activated powders at 1250°C and 25 MPa for 2 h [30]. However, the maximum permittivity of the ceramics prepared from single calcined powder was still about 2500. Just by using a double calcination, a dielectric permittivity up to 4080 and a dissipation factor between 0.001 and 0.016 at 1 kHz were reported for these dense hot-pressed KT ceramics [30].

Figure 5a presents the temperature dependence of the dielectric losses, tan*δ* at 10 kHz for stoichiometric and nonstoichiometric ST ceramics. For stoichiometric ST, a strong loss peak is evident at 70–105 K for the frequency range 10^2–10^6 Hz. Such a peak observed also in nominally pure ST single crystals can be attributed to the slowing down of polar modes at unavoidable defects within ferroelastic domain walls [39, 40]. ST ceramics with Sr/Ti = 1.02 show the concomitant suppression of the loss peak in contrast to the ceramics with Ti excess that have a very weak effect on the dielectric loss behaviour of ST. The minimum dissipation factor values are listed in **Table 1**.

From the temperature dependence of the dielectric loss at 10 kHz for KT ceramics with initial K/Ta ratio = 1, shown in **Figure 5b**, up to five peaks around 27, 49, 62, 127, and 214 K, can be detected. For K/Ta ratio = 1.02, similar but less intense peaks (below 0.007) are observed as well. Moreover, another low intense peak emerges at about 89 K. For K/Ta ratio = 1.05, the later peak grows, becoming

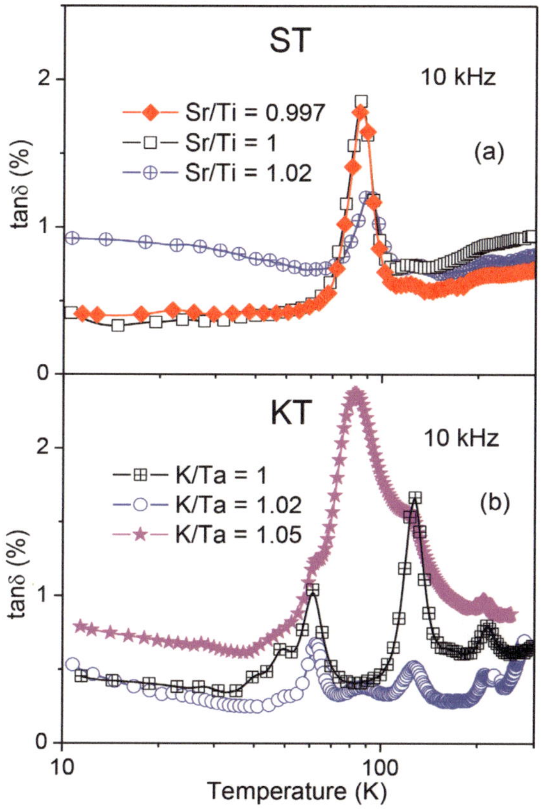

Figure 5.
Temperature dependence of the dissipation factor tanδ of strontium titanate ($SrTiO_3$, ST) (a) and potassium tantalate ($KTaO_3$, KT) (b) ceramics with indicated Sr/Ti and K/Ta ratios at frequency of 10 kHz (adapted from [32, 33]).

a dominant one and shifting to 83 K. All the other peaks mostly transform into shoulders. Therefore, potassium excess first decreases the dielectric loss down to 0.0025 (see also **Table 1**), reducing the peak intensities, but then strongly increases the loss up to 0.0237, inducing a strong peak close to 83 K. Thus, even compared with double calcined hot-pressed KT ceramics [30], close permittivity and lower losses could be obtained by conventional method just using 2% excess of potassium.

4. Analysis and discussion

The temperature dependences of the dielectric permittivity for both ST ceramics with Sr/Ti = 0997, 1, 102 and KT ceramics with K/Ta = 1, 1.02, and 1.05 were fitted by Barrett's relation (Eq. (3)). As shown in **Figure 4**, the fitting curves match well the $\varepsilon'(T)$ data points. The fitted parameters of the Barrett's relation for these experimental data are indicated in **Table 1**. T_0 values of 32–35 K for ST ceramics and of 10–14 K for KT ceramics as well as T_1 values of 98–110 and 48–66 K for ST and KT ceramics, respectively, are in agreement with those for corresponding ceramics and single crystals reported in the literature and summarised in **Table 2**.

As also seen from **Table 1**, whereas nonstoichiometry does not tend to change the transition temperature T_0 both for ST and KT, C parameter increases from 87×10^3 to 112×10^3 K with decreasing Sr/Ti ratio in ST and from 38×10^3 to

Composition	Type	T_0, K	T_1, K	$C/10^3$, K	ε_1	Ref.
$SrTiO_3$	SC	35.5	80.0	80.0	—	[9]
	CER	25.0	84.0	81.0	—	[37]
	CER	43.2	110.0	78.0	—	[38]
$KTaO_3$	SC	12.9	54.2	55.8	—	[28]
	SC	13.1	56.9	54.5	47.5	[24]
	SC	8.0	48.3	61.8	48.0	[24]
	CER	5.2	60.0	48.6	49.0	[24]
	CER	15.0	56.0	51.0	64.0	[30]

SC, single crystals; CER, ceramics.

Table 2.
Parameters of the Barrett's relation reported in literature for the quantum paraelectrics, $SrTiO_3$, ST and $KTaO_3$, KT.

57×10^3 K with increasing initial K/Ta ratio in KT ceramics. The increase of C reflects the increase of the $\varepsilon'(T)$ amplitude. In the case of KT, such increase might be explained by two reasons. On one hand, there is a strong increase of the grain size and decrease of the weak grain boundary contribution to the dielectric permittivity for K/Ta > 1. On the other hand, there is a reduction and final disappearance of the less polarisable potassium-poor secondary phase with increasing K/Ta ratio. In the case of ST, the increase of permittivity for Ti-rich ST can be related to the enhanced polarisability of the lattice, fully packed by Ti^{4+} ions. Ti^{4+} ions are the most polarisable in ST [41], while their collective off-central displacement is responsible for ferroelectricity establishment and therefore permittivity increase in titanates. On the other hand, the excess of Sr fully accommodates in the perovskite ST lattice forming interlayers within the RP structure. Such interlayers are expected to be less polarisable than the perovskite lattice [42], thus contributing to the lowering of the dielectric permittivity observed in the present work for Sr-rich ST. The reduction of the grain size in these compositions is another factor that can also contribute to such a decrease [36].

5. Conclusions

The effect of Sr/Ti ratio (0.997–1.02) and initial K/Ta ratio (1–1.05) on the phase morphology and dielectric response of ST and KT ceramics, respectively, is overviewed. Whereas no second phases were detected for the studied ST ceramics, initial excess of potassium was shown to be necessary to yield single-phase KT ceramics by solid state reaction process. Moreover, potassium excess favours the grain growth in KT ceramics, whereas Sr excess impedes the grain growth in ST ceramics and thus decreases the dielectric permittivity. On the contrary, Ti excess promotes the increase of the dielectric permittivity values of ST ceramics. Combination of the absence of secondary phases with increased grain size in KT ceramics with initial potassium excess results simultaneously in the increase of the lowest temperature dielectric permittivity value. Furthermore, the variation of Sr/Ti and K/Ta ratios did not change the quantum-paraelectric nature of ST and KT, respectively. Fitting the Barrett's relation to the experimental data revealed just considerable dissimilarities in the Curie-Weiss constants in agreement with the highest permittivity variation with A/B ratios, while characteristic temperatures did not change significantly.

Acknowledgements

This work was developed within the scope of the project CICECO-Aveiro Institute of Materials, FCT Ref. UID/CTM/50011/2019, financed by national funds through the FCT/MCTES.

Author details

Alexander Tkach and Paula M. Vilarinho*
Department of Materials and Ceramic Engineering, CICECO—Aveiro Institute of Materials, University of Aveiro, Aveiro, Portugal

*Address all correspondence to: paula.vilarinho@ua.pt

References

[1] Vilarinho PM. Functional materials: Properties, processing and applications. In: Vilarinho PM, Rosenwaks Y, Kingon A, editors. Scanning Probe Microscopy: Characterization, Nanofabrication and Device Application of Functional Materials. Dordrecht: Kluwer Academic Publishers; 2005. pp. 3-33

[2] Martin LW, Rappe AM. Thin-film ferroelectric materials and their applications. Nature Reviews Materials. 2016;**2**:16087

[3] Smolenskii GA, editor. Ferroelectrics and Related Materials. New York: Gordon and Breach Science Publishers; 1984

[4] Tagantsev AK, Sherman VO, Astafiev KF, Venkatesh J, Setter N. Ferroelectric materials for microwave tunable applications. Journal of Electroceramics. 2003;**11**:5-66

[5] Gevorgian S. Introduction: Overview of agile microwave technologies. In: Gevorgian S, editor. Ferroelectrics in Microwave Devices, Circuits and Systems. London: Springer Publishing; 2009. pp. 1-19

[6] Vendik OG, Ter-Martirisian LT, Zubko SP. Microwave losses in incipient ferroelectrics as functions of the temperature and the biasing field. Journal of Applied Physics. 1998;**84**:993-998

[7] Geyer RG, Riddle B, Krupka J, Boatner LA. Microwave dielectric properties of single-crystal quantum paraelectrics $KTaO_3$ and $SrTiO_3$ at cryogenic temperatures. Journal of Applied Physics. 2005;**97**:104111

[8] Lemanov VV, Sotnikov AV, Smirnova EP, Weihnacht M, Kunze R. Perovskite $CaTiO_3$ as an incipient ferroelectric. Solid State Communications. 1999;**110**:611-614

[9] Müller KA, Burkard H. $SrTiO_3$: An intrinsic quantum paraelectric below 4 K. Physical Review B. 1979;**19**:3593-3602

[10] Samara GA. The relaxational properties of compositionally disordered ABO_3 perovskites. Journal of Physics: Condensed Matter. 2003;**15**:R367-R411

[11] Barrett JH. Dielectric constant in perovskite type crystals. Physics Review. 1952;**86**:118-120

[12] Fleury PA, Scott JF, Worlock JM. Soft phonon modes and the 110°K phase transition in $SrTiO_3$. Physical Review Letters. 1968;**21**:16-19

[13] Uwe H, Sakudo T. Stress-induced ferroelectricity and soft phonon modes in $SrTiO_3$. Physical Review B. 1976;**13**:271-286

[14] Lemanov VV. Phase transitions in $SrTiO_3$ quantum paraelectric with impurities. Ferroelectrics. 1999;**226**:133-146

[15] Tkach A, Vilarinho PM, Kholkin AL. Polar behavior in Mn-doped $SrTiO_3$ ceramics. Applied Physics Letters. 2005;**86**:172902

[16] Levin I, Krayzman V, Woicik JC, Tkach A, Vilarinho PM. X-ray absorption fine structure studies of Mn coordination in doped perovskite $SrTiO_3$. Applied Physics Letters. 2010;**96**:052904

[17] Itoh M, Wang R, Inaguma Y, Yamaguchi T, Shan Y-J, Nakamura T. Ferroelectricity induced by oxygen isotope exchange in strontium titanate perovskite. Physical Review Letters. 1999;**82**:3540-3543

[18] Bhalla AS, Guo RY, Roy R. The perovskite structure—A review of its

role in ceramic science and technology. Materials Research Innovations. 2000;**4**:3-26

[19] Goldschmidt VM. Geochemisce verterlungsgesetze der elemente. Oslo: Norske Videnskap; 1927

[20] Witek S, Smyth DM, Pickup H. Variability of the Sr/Ti ratio in $SrTiO_3$. Journal of the American Ceramic Society. 1984;**67**:372-375

[21] Balachandran U, Eror NG. On the defect structure of strontium-titanate with excess SrO. Journal of Materials Science. 1982;**17**:2133-2140

[22] Wang Z, Caon M, Yaon Z, Li G, Song Z, Hu W, et al. Effects of Sr/Ti ratio on the microstructure and energy storage properties of nonstoichiometric $SrTiO_3$ ceramics. Ceramics International. 2014;**40**:929-933

[23] Amaral L, Tkach A, Vilarinho PM, Senos AMR. New insights into the effect of nonstoichiometry on the electric response of strontium titanate ceramics. Journal of Physical Chemistry C. 2019;**123**:710-718

[24] Chen ZX, Zhang XL, Cross LE. Low-temperature dielectric properties of ceramic potassium tantalate ($KTaO_3$). Journal of the American Ceramic Society. 1983;**66**:511-515

[25] Axelsson A-K, Pan Y, Valant M, Alford N. Synthesis, sintering and microwave dielectric properties of $KTaO_3$ ceramics. Journal of the American Ceramic Society. 2009;**92**:1773-1778

[26] Davis TG. Dielectric properties and soft modes in the ferroelectric mixed crystals $K_{1-x}Na_xTaO_3$. Physical Review B. 1972;**5**:2530-2537

[27] Samara GA, Morosin B. Anharmonic effects in $KTaO_3$-ferroelectric mode, thermal-expansion, and compressibility. Physical Review B. 1973;**8**:1256-1264

[28] Vogt H, Uwe H. Hyper-Raman scattering from the incipient ferroelectric $KTaO_3$. Physical Review B. 1984;**29**:1030-1034

[29] Zlotnik S, Vilarinho PM, Costa MEV, Moreira JA, Almeida A. Growth of incipient ferroelectric $KTaO_3$ single crystals by a modified self-flux solution method. Crystal Growth and Design. 2010;**10**(8):3397-3404

[30] Glinšek S, Malič B, Rojac T, Filipič C, Budič B, Kosec M. $KTaO_3$ ceramics prepared by the mechanochemically activated solid-state synthesis. Journal of the American Ceramic Society. 2011;**94**:1368-1373

[31] Zlotnik S, Sahu SK, Navrotsky A, Vilarinho PM. Pyrochlore and perovskite potassium tantalate: Enthalpies of formation and phase transformation. Chemistry—A European Journal. 2015;**21**:5231-5237

[32] Tkach A, Vilarinho PM, Senos AMR, Kholkin AL. Effect of nonstoichiometry on the microstructure and dielectric properties of strontium titanate ceramics. Journal of the European Ceramic Society. 2005;**25**:2769-2772

[33] Tkach A, Vilarinho PM, Almeida A. Role of initial potassium excess on the properties of potassium tantalate ceramics. Journal of the European Ceramic Society. 2011;**31**:2303-2308

[34] Levin EM, Robbins CR, McMurdie HF. Phase Diagrams for Ceramists. Columbus: The American Ceramic Society; 1964

[35] Reisman A, Holtzberg F, Berkenblit M, Berry M. Reactions of the group VB pentoxides with alkali

oxides and carbonates. III. Thermal and X-ray phase diagrams of the system K_2O or K_2CO_3 with Ta_2O_5. Journal of the American Chemical Society. 1956;**78**:4514-4520

[36] Petzelt J, Ostapchuk T, Gregora I, Rychetský I, Hoffmann-Eifert S, Pronin AV, et al. Dielectric, infrared, and Raman response of undoped $SrTiO_3$ ceramics: Evidence of polar grain boundaries. Physical Review B. 2001;**64**:184111

[37] Yu Z, Ang C. Dielectric and conduction behavior of La-doped $SrTiO_3$ with suppressed quantum-paraelectric background. Applied Physics Letters. 2002;**80**:643-645

[38] Tkach A, Okhay O, Vilarinho PM, Kholkin AL. High dielectric constant and tunability of strontium titanate ceramics modified by chromium doping. Journal of Physics: Condensed Matter. 2008;**20**:415224

[39] Viana R, Lunkenheimer P, Hemberger J, Bohmer R, Loidl A. Dielectric spectroscopy in $SrTiO_3$. Physical Review B. 1994;**50**:601-604

[40] Ang C, Guo R, Bhalla AS, Cross LE. Effect of electric field and post-treatment on dielectric behavior of $SrTiO_3$ single crystal. Journal of Applied Physics. 2000;**87**:3937-3940

[41] Shannon RD. Dielectric polarizabilities of ions in oxides and fluorides. Journal of Applied Physics. 1993;**73**:348-366

[42] Wise PL, Reaney IM, Lee WE, Iddles DM, Cannell DS, Price TJ. Tunability of τ_f in perovskites and related compounds. Journal of Materials Research. 2002;**17**:2033-2040

Chapter 5

Thermodynamic Stability and Microscopic Behavior of $Ba_xSr_{1-x}Co_{1-y}Fe_yO_{3-\delta}$ Perovskites

Florentina Maxim, Alina Botea-Petcu, Florina Teodorescu, Ludwig J. Gauckler and Speranta Tanasescu

Abstract

The mixed conducting perovskite-type oxides $Ba_xSr_{1-x}Co_{1-y}Fe_yO_{3-\delta}$ (BSCF) are intensively studied as potential high-performance solid oxide fuel cell cathode materials. The effect of different compositional variables and oxygen stoichiometry on the structure and thermodynamic stability of the $Ba_xSr_{1-x}Co_{1-y}Fe_yO_{3-\delta}$ (x = 0.2, 0.4, 0.5, 0.6, 0.8; y = 0.2, 0.4, 0.6, 0.8, 1) perovskite-type compositions were investigated by solid electrolyte electrochemical cells method and scanning electron microscopy (SEM). The thermodynamic quantities represented by the partial molar free energies, enthalpies and entropies of oxygen dissolution in the perovskite phase, as well as the equilibrium partial pressures of oxygen were obtained in the temperature range of 823–1273 K. The *in situ* change of oxygen stoichiometry and the determination of thermodynamic parameters of the new oxygen-deficient BSCF compositions were studied *via* coulometric titration technique coupled with electromotive force (EMF) measurements. The effect of A- and B-site dopants concentration correlated to the variation of oxygen stoichiometry on the thermodynamic stability and morphology of the BSCF samples was evidenced.

Keywords: BSCF, perovskite-type compounds, oxygen stoichiometry, thermodynamic data, electromotive force measurements, scanning electron microscopy, cathodes SOFC

1. Introduction

The series $Ba_xSr_{1-x}Co_{1-y}Fe_yO_{3-\delta}$ (BSCF) perovskites are well known for their good oxygen catalytic activity and mixed ionic-electronic conductivity (MIEC) and gained attention as promising electrode materials for solid oxide fuel cells (SOFCs) and oxygen-permeable membranes. Depending on temperature and oxygen partial pressure the BSCF perovskites exhibit high oxygen non-stoichiometry ($0.3 < \delta < 0.8$ at $873 < T < 1073$ K) [1–4]. This extraordinary ability to host oxygen vacancies and to transport a significant amount of oxygen anions *via* oxygen vacancies (with an oxygen diffusion coefficient of about 10^{-6} cm^2/s and ionic conductivity of 0.018 S/cm at 973 K) has been reported for $Ba_{0.5}Sr_{0.5}Co_{0.8}Fe_{0.2}O_{3-\delta}$ [4–8]. However, significant differences were observed in the structural characteristics and the electrochemical performance of the BSCF solid solutions as a function of temperature (T), atmosphere (inert [9], oxidizing [10–12] and reducing [13, 14]), as well as of the nature and concentration of A- and B-sites substituents [15–18]. As consequence, there are questions regarding the long-term stability and performance of such highly defective material under the operating conditions of an intermediate temperature solid oxide fuel cell (IT-SOFC) of 873–1073 K. At low oxygen partial pressures (under 10^{-5} atm) and in reducing conditions the structural changes through different pathways was shown [16, 17, 19]; the variation of the oxidation state of the Co ion was suggested as the driving force for these changes [14]. The formation of non-cubic phases might limit the performance of the oxygen ion conduction, by increasing the lattice stress and decreasing the oxygen vacancies mobility.

Despite the interest and the research effort in this field, many aspects of finding appropriate processing parameters and, above all, the fundamental understanding of the correlations between all the factors that ensure the optimization of the SOFC cathodes are not yet elucidated.

The aim of the study is to evidence the effect of the composition, and oxygen stoichiometry change on the thermodynamic properties and morphology of perovskite-type oxides in the $Ba_xSr_{1-x}Co_{1-y}Fe_yO_{3-\delta}$ (x = 0.2, 0,4, 0.5, 0,6, 0.8; y = 0.2, 0.4, 0.6, 0.8 and 1) (BSCF) system. In addition, the correlative effect of temperature and defect structure on the thermodynamic behavior of the samples was discussed based on the evolution of the thermodynamic quantities of the oxygen dissolution in the perovskite structure. The partial molar thermodynamic properties of oxygen dissolution in the perovskite phase, as well the equilibrium partial pressures of oxygen have been obtained by using solid electrolyte electrochemical cells method. The influence of the oxygen stoichiometry change on the thermodynamic properties was examined using the data obtained by a coulometric titration technique coupled with EMF measurements.

2. Materials and methods

The details of the sample preparation method are presented elsewhere [13]. Briefly, powder specimens of BSCF were obtained by solid state reaction starting from barium carbonate, strontium carbonate, iron oxide and cobalt oxide raw materials. In order to reach the phase equilibrium of the desired perovskite, the powder mixture was ground and calcined for several times at 1273 K for 10 h. The X-ray diffraction analysis of the as prepared powder samples (shown elsewhere [13]) demonstrates the formation of a predominant cubic phase for all the BSCF investigated compositions, although small amounts of hexagonal phase could be present in the BSCF 8282 sample [20, 21].

The morphology of BSCF powders was analyzed by SEM using a FEI Quanta 3D equipment operated at low acceleration voltage (maximum 5 kV) and using the backscatter detector in beam deceleration mode. This SEM mode enables high resolution imaging and high surface sensitivity [22], and it was used in this study to evidence the surface features of the BSCF particles.

The electrochemical cell method was employed to obtain the thermodynamic properties of the BSCF samples. The experimental setups as well as theoretical aspects were described in detail in previous papers [23–25]. The electrochemical cell contains a yttria stabilized zirconia solid electrolyte and an iron- wüstite reference electrode:

(−) Fe, wustite/ZrO_2 (Y_2O_3) /BSCF (+)

where BSCF denotes the $Ba_xSr_{1-x}Co_{1-y}Fe_yO_{3-\delta}$ (x = 0.2, 0.4, 0.5, 0.6, 0.8; y = 0, 0.2, 0.4, 0.6, 0.8, 1) perovskite-type samples.

Measurements were performed in vacuum at a residual gas pressure of 10^{-5} atm. The EMF was measured with a Keithley 2000 multimeter, at 50 K intervals between

823 and 1273 K, each time waiting until equilibrium conditions were obtained. Equilibrium conditions were achieved when EMF values for increasing and decreasing temperatures agreed within ±1 mV within a-30 minutes interval.

The solid-state coulometric titration technique was used to accurately change the oxygen stoichiometry of BSCF pellets. The titrations were performed *in situ*, in vacuum at 1123 K by using a Bi-PAD Tacussel potentiostat. The mass change $|\Delta m|$ (g) of the sample is associated to the transferred charge $Q(A \cdot s)$, in agreement to Faraday's law.

$$|\Delta m| = 8.291 \bullet 10^{-5}\, Q \tag{1}$$

For every new composition obtained, the equilibrium EMF's values at 50 K intervals between 1073 and 1273 K are recorded in the open circuit condition.

3. Results and discussions

The BSCF samples with different cation compositions analyzed in this study are listed in **Table 1**. In order to study the effect of the A- and B- site composition of the perovskite structure on the thermodynamic properties and particles' morphologies, the samples were grouped in two sets corresponding to the variation of Ba (x) and Fe (y) concentration, respectively.

3.1 A-site effect: Ba/Sr variation

The micrographs of the as prepared BSCF powders obtained for increasing Ba content, a) x = 0.2, b) x = 0.4, c) x = 0.5, d) x = 0.6 and e) x = 0.8 are presented in **Figure 1**. There are two categories of morphologies specific to different compositions analyzed in this study. The first type is the well-defined particles with similar shape and size (2–5 μm) that are formed in powders with the lowest and the highest concentration of Ba (BSCF 2882 - **Figure 1a** and BSCF 8282 - **Figure 1e**, respectively). The second category of morphology is represented by particles with round tip branches and with nanoscale features at their surface. Particles with such surface characteristics are observed for the BSCF 4682, BSCF 5582, and BSCF 6482 samples (**Figure 1b–d**). However, these particles are different in size and in shape depending on the Ba content. The powders with the BSCF 4682 (**Figure 1b**) and BSCF 6482 (**Figure 1d**) compositions have particles with dimensions exceeding 10 μm in length, while particles with 50% of Ba are around 3 μm in size (**Figure 1c** – BSCF 5582). Moreover, there are also differences in the shape of these surface microstructures. The particles with BSCF 4682 composition have spherical features on the surface (**Figure 1b**), while for the BSCF 6482 sample (**Figure 1d**) the nanoscale surface structures are elongated and the tendency to agglomerate at the

$Ba_xSr_{1-x}Co_8Fe_2O_{3-\delta}$	Abbreviation	$Ba_{0.5}Sr_{0.5}Co_{1-y}Fe_yO_{3-\delta}$	Abbreviation
$Ba_{0.2}Sr_{0.8}Co_{0.8}Fe_{0.2}O_{3-\delta}$	BSCF 2882	$Ba_{0.5}Sr_{0.5}Co_{0.8}Fe_{0.2}O_{3-\delta}$	BSCF 5582
$Ba_{0.4}Sr_{0.6}Co_{0.8}Fe_{0.2}O_{3-\delta}$	BSCF 4682	$Ba_{0.5}Sr_{0.5}Co_{0.6}Fe_{0.4}O_{3-\delta}$	BSCF 5564
$Ba_{0.5}Sr_{0.5}Co_{0.8}Fe_{0.2}O_{3-\delta}$	BSCF 5582	$Ba_{0.5}Sr_{0.5}Co_{0.4}Fe_{0.6}O_{3-\delta}$	BSCF 5546
$Ba_{0.6}Sr_{0.4}Co_{0.8}Fe_{0.2}O_{3-\delta}$	BSCF 6482	$Ba_{0.5}Sr_{0.5}Co_{0.2}Fe_{0.8}O_{3-\delta}$	BSCF 5528
$Ba_{0.8}Sr_{0.2}Co_{0.8}Fe_{0.2}O_{3-\delta}$	BSCF 8282	$Ba_{0.5}Sr_{0.5}FeO_{3-\delta}$	BSCF 5501

Table 1.
The BSCF compositions analyzed in this study and their corresponding notations.

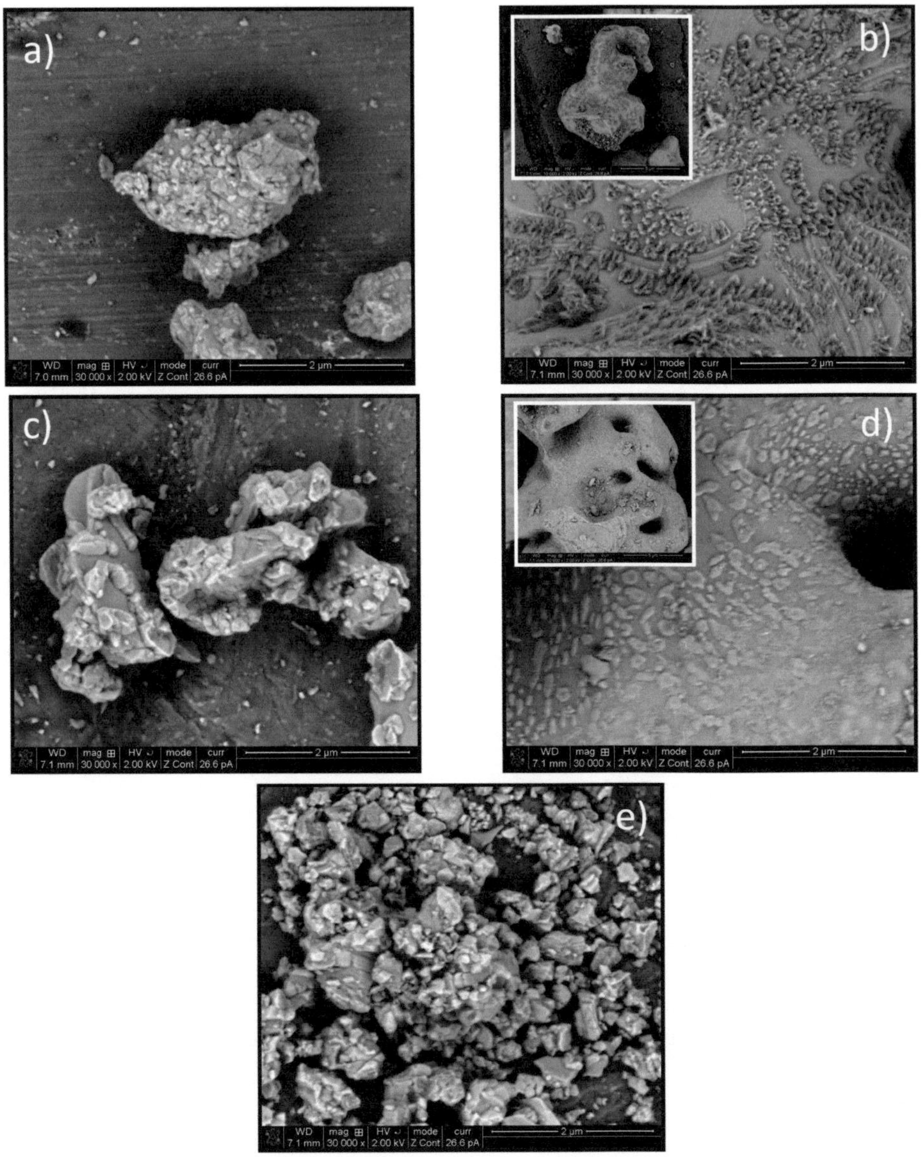

Figure 1.
SEM micrographs of the BSCF powders with increasing Ba content (x): a) BSCF 2882, b) BSCF 4882, c) BSCF 5582, d) BSCF 6482, e) BSCF 8282; scale bar for the main images is 2 µm, while for the inset in b) and d) is 5 µm.

particle tip. The peculiar nanoscaled features on the BSCF particles surface were also observed by other authors for powders prepared by coprecipitation [26] and by solution combustion synthesis [27], and further calcined at temperatures as high as 1273 K. It was claimed that the surface nanostructures induced the high oxygen conductivity of BSCF powders [26]. Another postulated hypothesis is that these nanoscaled features on the surface of BSCF powders appear due to minor amounts of oriented hexagonal phase present as secondary phase [27, 28]. However, BSCF powders prepared by the solid state reaction method used in this study are expected to have a predominantly cubic perovskite structure [20].

The variation of partial molar free energy ($\Delta\bar{G}_{O_2}$) and oxygen partial pressure (p_{O_2}) with temperature is presented in **Figure 2** for the five selected BSCF compositions corresponding to different Ba/Sr ratios with x ranging from 0.2 to 0.8.

The variation of $\Delta\bar{G}_{O_2}$ and log p_{O2} with temperature is not monotonous. In the temperature range of 823–923 K, the BSCF 5582 and BSCF 2882 exhibited the highest values of partial molar free energy, while in the domain 923–1223 K, the highest partial molar free energy values were noted for BSCF 6482. For instance, in this temperature range, $\Delta\bar{G}_{O_2}$ value for BSCF 6482 is up to 30 kJ mol^{-1} higher compared to that of the BSCF 5582 sample. BSCF 5582 exhibited the flattest variation of $\Delta\bar{G}_{O_2}$ with temperature in the entire temperature domain 823–1273 K. The highest oxygen vacancy concentration is expected for the BSCF 6482 composition, as high $\Delta\bar{G}_{O_2}$ values were recorded. A higher oxygen vacancy concentration means more carriers for oxygen transport. Therefore, a higher oxygen ionic conductivity is envisaged for BSCF 6482 above 923 K. This statement is consistent with the electrical conductivity and improved electrochemical performance reported for the BSCF 6482-based cathode material at these temperatures [29]. High oxygen conductivity and oxygen permeability at intermediate temperatures were also reported for the BSCF 5582 composition [16, 30]. Doping with 50–60% Ba stabilizes the lower oxidation states of the B-site cations, holding the perovskite-structure more effectively and thus contributing indirectly to the enhanced electrochemical performance for these compositions [31, 32].

To get insights into the energetics of oxygen vacancy formation, the partial molar enthalpy and entropy of oxygen dissolution in the perovskite lattice ($\Delta\bar{H}_{O_2}$ and $\Delta\bar{S}_{O_2}$, respectively) were calculated in the temperature domains in which the partial molar free energies are linear functions of temperature. For each sample, these particular temperature domains are: 973–1123 K and 1123–1223 K for BSCF 2882; 973–1073 K and 1073–1173 K for BSCF 4682; 973–1073 K and 1073–1223 K for BSCF 5582; 923–1023 K and 1023–1173 K for BSCF 6482; 923–1023 K and 1023–1123 K for BSCF 8282 (**Figure 3a**). The thermodynamic investigation pointed out that the temperature of structural transformations (T_{tr}) decreases as Ba (x) content increases, being 1123 K for x = 0.2; 1073 K for x = 0.4 and x = 0.5; 1023 K for x = 0.6 and x = 0.8.

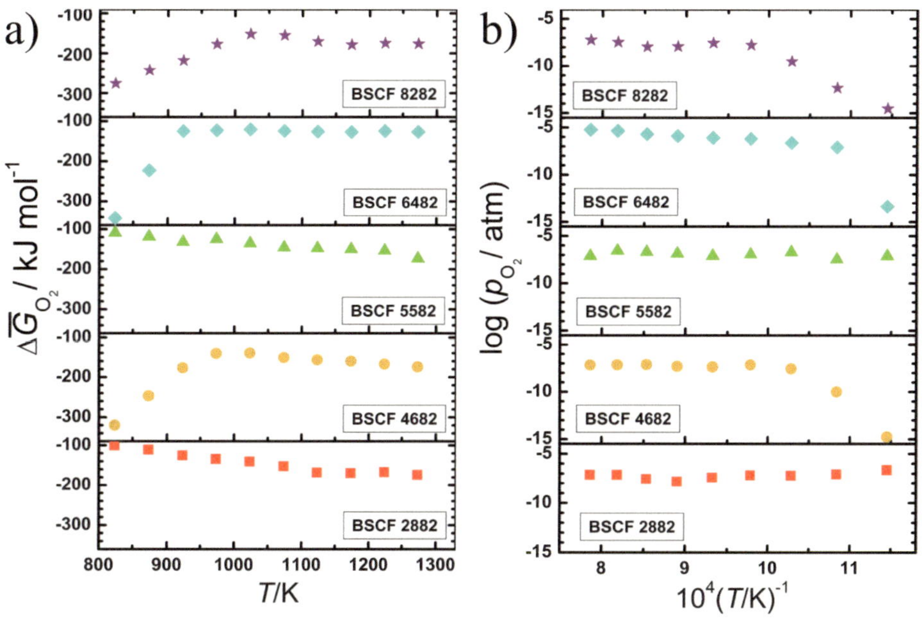

Figure 2.
Variation of a) $\Delta\bar{G}_{O_2}$ and b) log p_{O_2} with the Temperature and Ba-content (x).

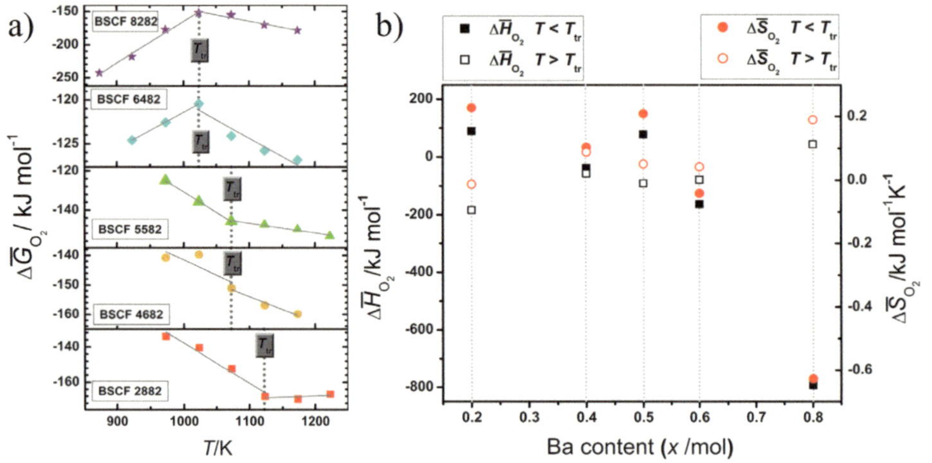

Figure 3.
a) Variation of $\Delta\bar{G}_{O_2}$ with temperature and Ba content (x) - linear fit in the selected temperature domain; b) $\Delta\bar{H}_{O_2}$ and $\Delta\bar{S}_{O_2}$ versus *Ba content (x) before and after T_{tr} defined in a).*

The variation of $\Delta\bar{H}_{O_2}$ and $\Delta\bar{S}_{O_2}$ with Ba content is depicted in **Figure 3b**. At temperatures lower than T_{tr}, the $\Delta\bar{H}_{O_2}$ and $\Delta\bar{S}_{O_2}$ for the samples with 20, 40 and 50% of Ba are in the similar range. As the Ba concentration increases, the partial molar enthalpy and entropy values drastically decrease, reaching values as low as −792 kJ mol^{-1} and -627 J mol^{-1} K^{-1}, respectively, for the BSCF 8282 composition.

This decrease of $\Delta\bar{H}_{O_2}$ and $\Delta\bar{S}_{O_2}$ is associated with an increase of the binding energy of oxygen and an increase of order in the oxygen sublattice of the perovskite-type structure, respectively. At temperatures higher than the points of transitions, both $\Delta\bar{H}_{O_2}$ and $\Delta\bar{S}_{O_2}$ increased as the Ba-content increases. For BSCF 5582 and BSCF 6482 specimens these values are similar. Thus, besides the oxygen vacancy concentration, the ordering of the oxygen vacancy has a remarkable influence on the entropy values being an indication that the oxygen vacancies are distributed randomly on the oxygen sublattice.

In the intermediate temperature range below 1173 K the variations of the thermodynamic data show some anomalies, which could be correlated with the transition to the cubic BSCF structures [32]. These structural transformations are connected to the charge compensation mechanism. Crystal structure and electrical conductivity of several selected compositions in the Ba–Sr–Co–Fe–O system indicate that doping with more Ba into the system increases the ability for lattice oxygen exchange [13, 20, 21, 33]. A reversible phase transition from cubic to mixed phase of cubic and hexagonal at 973–1173 K for the BSCF 5582 compositions was pointed out both experimentally (employing coulometric titrations and thermal analysis [14, 20, 21, 32–35] and theoretically, by applying the density functional theory calculations [36, 37].

Keeping in mind the key role of oxygen vacancy ordering on the crystalline phase formation, the less symmetrical non-cubic phases are expected to have highly ordered oxygen vacancies. When the temperature increases above 1023 K, the vacancy ordering starts to disappear, the oxygen vacancies become more mobile and the crystalline phase of the material tends to exhibit higher symmetry, but lower stability. Both, the thermodynamic data and the phase symmetry results let us conclude that low symmetric BSCF perovskites, like BSCF 5582, BSCF 6482 and BSCF 4682 are thermodynamically more stable than the high symmetric BSCF perovskite.

3.2 B-site effect: Co/Fe variation

The morphological evolution of the as prepared BSCF powders with increasing Fe content, is shown in **Figure 4** for a) $y = 0.2$, b) $y = 0.4$, c) $y = 0.6$, d) $y = 0.8$, and e) $y = 1.0$. Well defined nanoscale features are shown in **Figure 4a** for the BSCF 5582 sample (see also **Figure 1c**). In contrast, smooth surfaces formed on the particles in the BSCF 5546 and the BSCF 5564 compositions (**Figure 4b** and **c**). The BSCF 5528 and the BSCF 5501 powders have a rough surface showing the incipient formation of the nanostructured features on the surface (**Figure 4d** and **e**). It is therefore expected that these samples to contain a very small quantity of the hexagonal secondary phase [27], not detectable by X-ray diffraction [20].

In **Figure 5**, the variation with temperature of the partial molar free energy (a) and of the oxygen partial pressure (b) is shown for the BSCF compositions with variable Fe content ($0.2 \leq y \leq 1$).

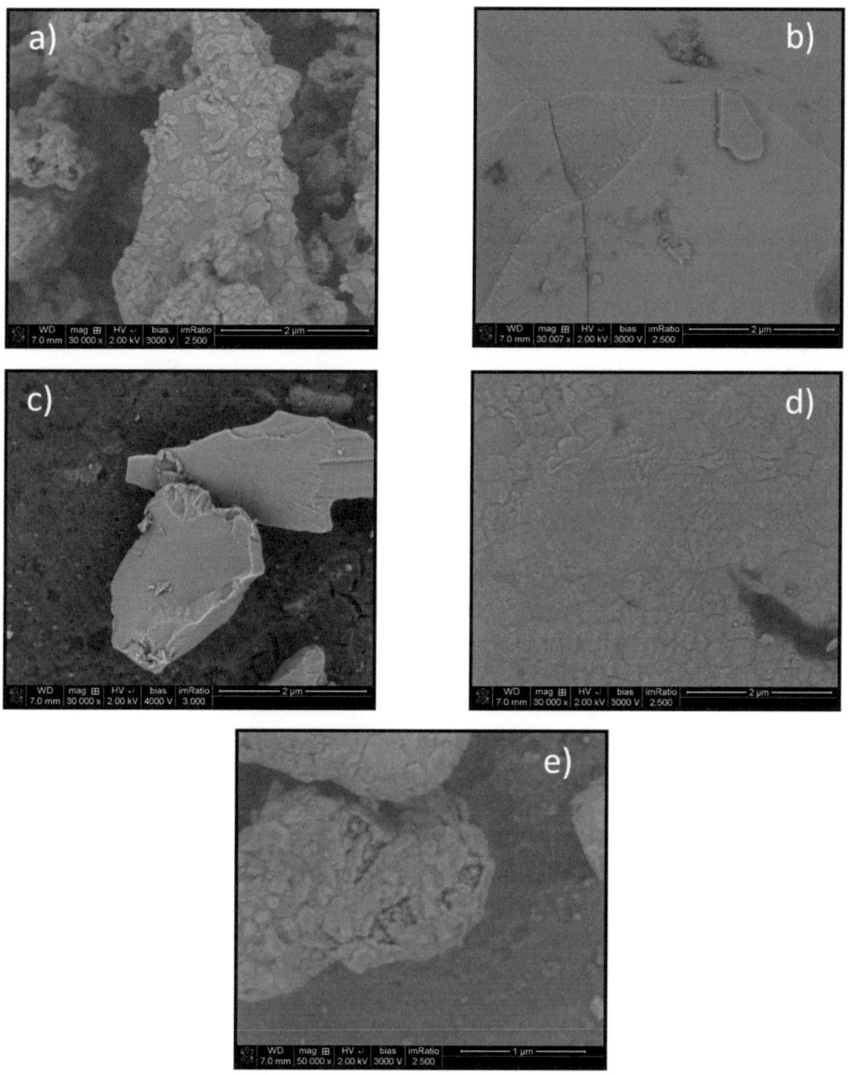

Figure 4.
SEM micrographs of the BSCF powders with increasing Fe content (y): a) BSCF 5582, b) BSCF 5564, c) BSCF 5546, d) BSCF 5528, e) BSCF 5501; scale bar for the images is 2 μm, except for e) where it is 1 μm.

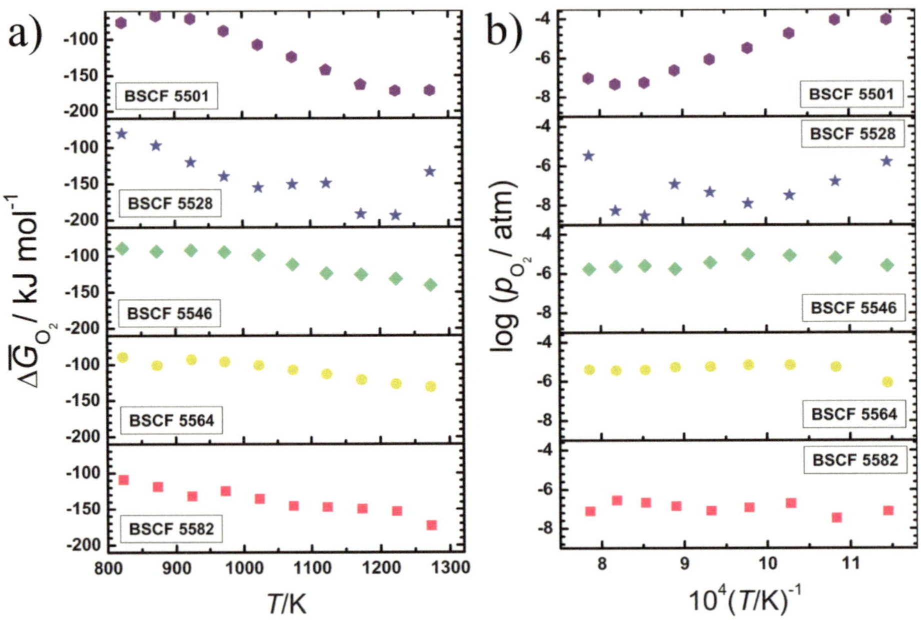

Figure 5.
a) $\Delta\bar{G}_{O_2}$ and b) log p_{O_2} with the temperature and Fe content (y).

The heated samples exhibited a complex behavior in the entire investigated temperature domain. One can observe that, in the intermediate temperature range from 823 K to 973 K, the BSCF 5501 sample has the highest recorded values for both $\Delta\bar{G}_{O_2}$ and *log* p_{O_2}, while, in the high temperature range from 1073 K to 1273 K, the highest values were obtained for the BSCF 5564 composition. This suggests the highest concentration of oxygen vacancies in the specified temperature ranges for these samples. The results could be correlated to the high oxygen non-stoichiometry [1], high ionic conductivity [38], as well as with the lattice expansion (volume effect) of the crystalline structure [39] reported for these compositions. For the BSCF 5582 it has been reported that about 60% of the Co^{3+} ions are oxidized to Co^{4+} ($r_{Co^{4+}} = 0.67$ Å), while, at the same time, all the Fe^{3+} ions are into high valence state (Fe^{4+} with $r_{Fe^{4+}} = 0.73$ Å), thus this effect has to be compensated by oxygen vacancy formation [40, 41]. Increasing iron content in the heat-treated samples at 1173 K and 1223 K, the decrease of the non-stoichiometry and of the total electrical conductivity is expected [1, 40]. This effect was also evidenced for various substituted ABO_3 perovskites solid solution series (e.g. $La_{1-x}Sr_xCo_{1-y}Fe_yO_{3-\delta}$ [42, 43] and $La_{1-x}Sr_xMn_{1-y}Fe_yO_{3-\delta}$ [44]) where the formation of oxygen vacancies decreased with Fe content.

The values for the partial molar enthalpy and entropy, $\Delta\bar{H}_{O_2}$ and $\Delta\bar{S}_{O_2}$, respectively, were calculated in the temperature ranges in which the partial molar free energies are linear functions of temperature (**Figure 5a**). These specific temperature domains are 973–1073 K and 1073–1223 K for BSCF 5582; 973–1273 K for BSCF 5564; 923–1023 K and 1023–1123 K for BSCF 5546 and 923–1173 K for BSCF 5501. The temperatures of transition (T_{tr}) were established at: 1073 K for y = 0.2, 1023 K for y = 0.6 and 1023 K for x = 0.8. The variation of $\Delta\bar{H}_{O_2}$ and $\Delta\bar{S}_{O_2}$ with Fe content, below and above T_{tr}. are depicted in **Figure 6b**. At temperatures below the transition temperature, $\Delta\bar{H}_{O_2}$ and $\Delta\bar{S}_{O_2}$ values of the BSCF compositions having 20, 40 and 60% Fe slightly decreased, and then increased again for the 80 and 100% of Fe concentrations. Our results correlate well with the parabolic variation in ionic

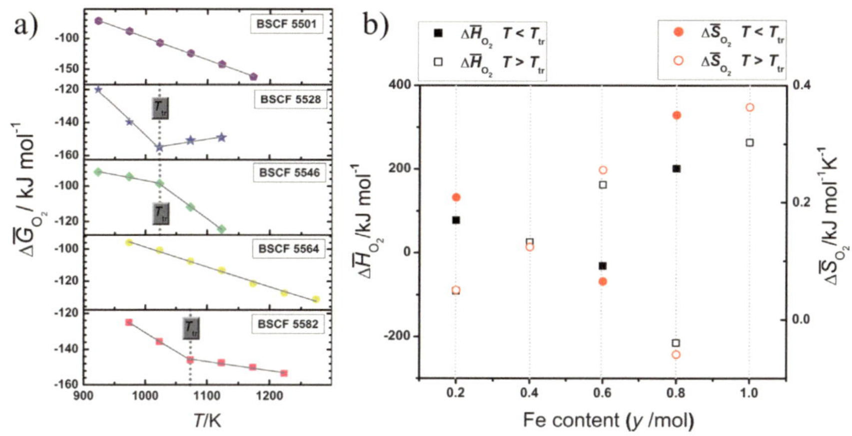

Figure 6.
a) Partial molar enthalpy $\Delta\bar{G}_{O_2}$ with temperature and Fe content (y) - linear fit in the selected temperature domains; b) partial molar enthalpy and entropy $\Delta\bar{H}_{O_2}$ and $\Delta\bar{S}_{O_2}$ as a function of Fe content (y) before and after T_{tr} defined in a).

conductivity *versus* iron content (with the identification of a minimum ionic conductivity at around 50% Fe-content) noted for $Ba_{1-x}Sr_xCo_{1-y}Fe_yO_{2.5}$ system [45]. This behavior is ascribed to the largest amount of oxygen vacancy trapped around Sr and Co ions for the composition containing ~50% iron. Above the T_{tr}, both $\Delta\bar{H}_{O_2}$ and $\Delta\bar{S}_{O_2}$ increased with the Fe-content increasing, except the values obtained for BSCF 5528. The increase of the partial molar enthalpy and entropy values with Fe content follows the order: BSCF 5582 < BSCF 5564 < BSCF 5546 < BSCF 5501 and suggests the decrease of the binding energy of oxygen in the lattice and a random distribution of the oxygen vacancies in the oxygen sublattice of the perovskite-type structure with the iron content. This mean that the thermodynamic stability of $Ba_{0.5}Sr_{0.5}Co_{1-y}Fe_yO_{3-\delta}$ increased in the following order $y = 1 < y = 0.6 < y = 0.4 < y = 0.2$, the specimen BSCF 5582 being the most stable composition for temperature $\geq$ 1023 K. This finding can be explained by the relative redox stability of the B^{3+} ions which seems to modify both the mobility and the concentration of the oxygen vacancies, at the same A-site composition. Moreover, the substitution of Co by Fe induces a stabilization of cubic perovskite structure [20, 46].

The BSCF 5528 sample has a distinct thermodynamic behavior which is further discussed. At temperatures lower than 1023 K, the values of enthalpy and entropy of 264 kJ mol^{-1} and 362 J mol^{-1} K^{-1}, respectively were obtained (**Figure 6b**). In the interval 1023–1123 K a strong decrease of the partial molar enthalpy and entropy was observed to values as low as −215 kJ mol^{-1} and -58.9 J mol^{-1} K^{-1}, respectively. In this temperature domain the BSCF 5528 sample exhibits high thermodynamic stability. Between 1173 and 1223 K, the variation of the partial molar free energy is observed (**Figure 5a**), which can be due to further structural transformation related to the charge compensation mechanism. The result is in accordance with the literature indicating the presence of secondary phases in the X-ray diffraction patterns of the samples following the thermal cycle at 1173 K [39]. At the same time, a sharp decrease in the permeation flux was reported for BSCF membranes with the increase of iron concentration from 60 to 80% [1]. The results suggest that the increase of iron concentration in BSCF might be hindered more by the slow oxygen bulk diffusion than by the surface exchange kinetics of the oxides. This could also explain the behavior of the BSCF 5528 sample at 1273 K for which the partial molar free energy increases with ~40 kJ mol^{-1} above the values corresponding to all the other investigated samples.

3.3 Oxygen non-stoichiometry effect

In order to further evaluate the previous results, the influence of the change of the oxygen stoichiometry on the thermodynamic properties was examined by solid state coulometric titration technique coupled with EMF measurements. The oxygen stoichiometry was modified by decreasing the stoichiometry with the same relative deviation of $\Delta\delta = 0.01$ for all the BSCF compositions. Further, the effect of the oxygen non-stoichiometry was correlated with the influence of the A- and B-site dopant.

Two sets of data representing the $\Delta\bar{G}_{O_2}$ change before and after the isothermal titration experiments, in the temperature range 1123–1273 K, for BSCF compositions are plotted in **Figure 7(a)** (for the samples with Ba-content increasing) and **Figure 7(b)** (for the samples with variable Fe concentration).

After titration, the decrease of $\Delta\bar{G}_{O_2}$ values with the stoichiometry change is observed for all the investigated compositions with various Ba- and Fe-dopant concentrations, except for the sample BSCF 5528 containing 80% Fe. The obtained results suggested that the energy of vacancy formation decreases with increasing non-stoichiometry while for BSCF 5528 an opposite behavior is estimated.

Considering the partial pressure of oxygen as a key parameter for the thermodynamic characterization of the materials, the variation of the $log\ p_{O_2}$ with the temperature and the concentration of the dopants at the same deviation from stoichiometry was analyzed. The results are presented in **Figure 8**.

A decrease of $log\ p_{O_2}$ with the relative non-stoichiometry change (δ) is observed for both series of BSCF compositions enriched in Ba (**Figure 8a**) and Fe (**Figure 8b**), excepting BSCF 5528 specimen. The obtained results for the BSCF analyzed compositions confirmed that at temperatures from 1123 K to 1273 K, the oxygen vacancies are generated at the expense of electron holes. The charge imbalance caused by the B-site substitution starts to be compensated by the formation of the oxygen vacancies. After oxygen coulometric titration, the highest deviation in the $log\ p_{O_2}$ values with the stoichiometry is obtained for the sample containing 100% of Fe (BSCF 5501). This, in turn, is accompanied by the highest decrease in the concentration of oxygen vacancies comparing to the other compositions.

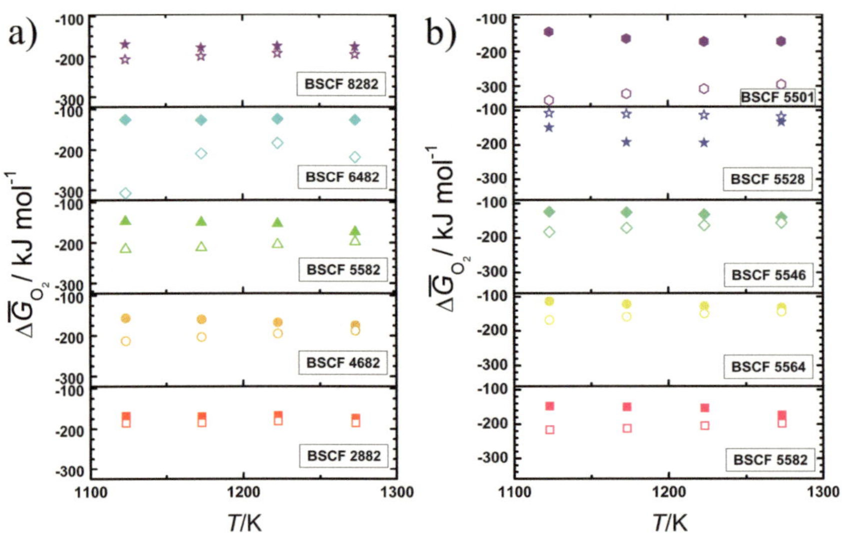

Figure 7.
$\Delta\bar{G}_{O_2}$ vs. *temperature for BSCF compounds with a) Ba-content* (x) *and b) Fe-content* (y) *variance showing the effect of the oxygen non-stoichiometry; solid symbols in each graph represent the values recorded before titration, while empty symbols indicate the corresponding values after titration.*

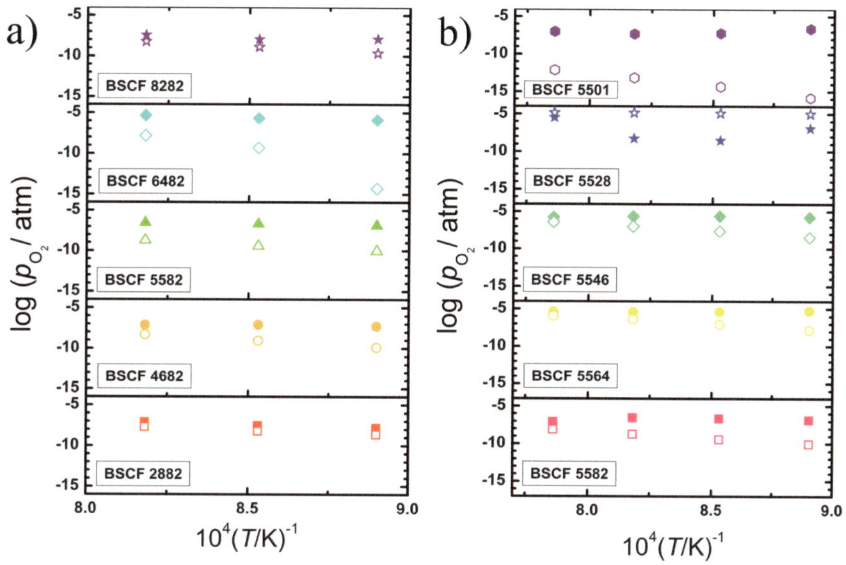

Figure 8.
Variation of log p_{O_2} with temperature for BSCF compounds with a) Ba-content (x) and b) Fe-content (y) variance showing the effect of the oxygen non-stoichiometry; solid symbols in each graph represent the values recorded before titration, while empty symbols indicate the corresponding values after titration.

The changes of $\Delta\bar{H}_{O_2}$ and $\Delta\bar{S}_{O_2}$ with the oxygen non-stoichiometry were evaluated for the BSCF compositions (**Figure 9**). For BSCF 6482 and BSCF 5528, the values of enthalpy and entropy were not included in **Figure 9a** and **9b**, respectively due to the deviation from linearity of the energy values obtained in the temperature domain 1123–1273 K. For both BSCF series with Ba- and Fe- compositional variations, the values of enthalpies and entropies decreased with stoichiometry change suggesting an increase in the binding energy of oxygen and change of order in the oxygen sublattice of the perovskite-type structure.

The $\Delta\bar{H}_{O_2}$ and $\Delta\bar{S}_{O_2}$ did not indicate a clear tendency of variation with barium concentration at the same deviation of stoichiometry, our results being consistent with the results of Girdauskaite *et al.* [47].

For the BSCF series with different Fe-content, a strong decrease of partial molar enthalpy with δ and with increasing of Fe-content was recorded, particularly for the sample with the highest Fe content. This result along with the negative enthalpy values indicated that the thermodynamic stability increased in the following order BSCF 5582 < BSCF 5564 < BSCF 5546 < BSCF 5501. Regarding the entropy, an abrupt decrease for the samples containing 20, 40 and 60% iron was observed, while for the composition containing only Fe in B-site, a slight $\Delta\bar{S}_{O_2}$ diminution with δ was noticed. The values of partial entropies of oxygen dissolution for y = 0.2, 0.4, and 0.6 are negative, which is an indicative for a metal vacancy mechanism [48, 49]. Due to the large decrease in $\Delta\bar{S}_{O_2}$ for the samples with y = 0.2; 0.4 and 0.6, it is considered that the oxygen vacancies would not randomly distribute on all of the oxygen sites but they would be distributed to some particular oxygen sites. This means that the total number of sites employed by oxygen vacancies is decreased. It is also possible that the vacancy distribution is related to some crystallographic distortions or ordering of metal sites. The result is in agreement with previous reports noting the formation of partially ordered oxygen vacancies in highly defective perovskite-type oxides [42, 50, 51].

The thermodynamic data showed that, after titration, BSCF 5501 is the most stable composition, exhibiting an increased binding energy of oxygen in the lattice but a

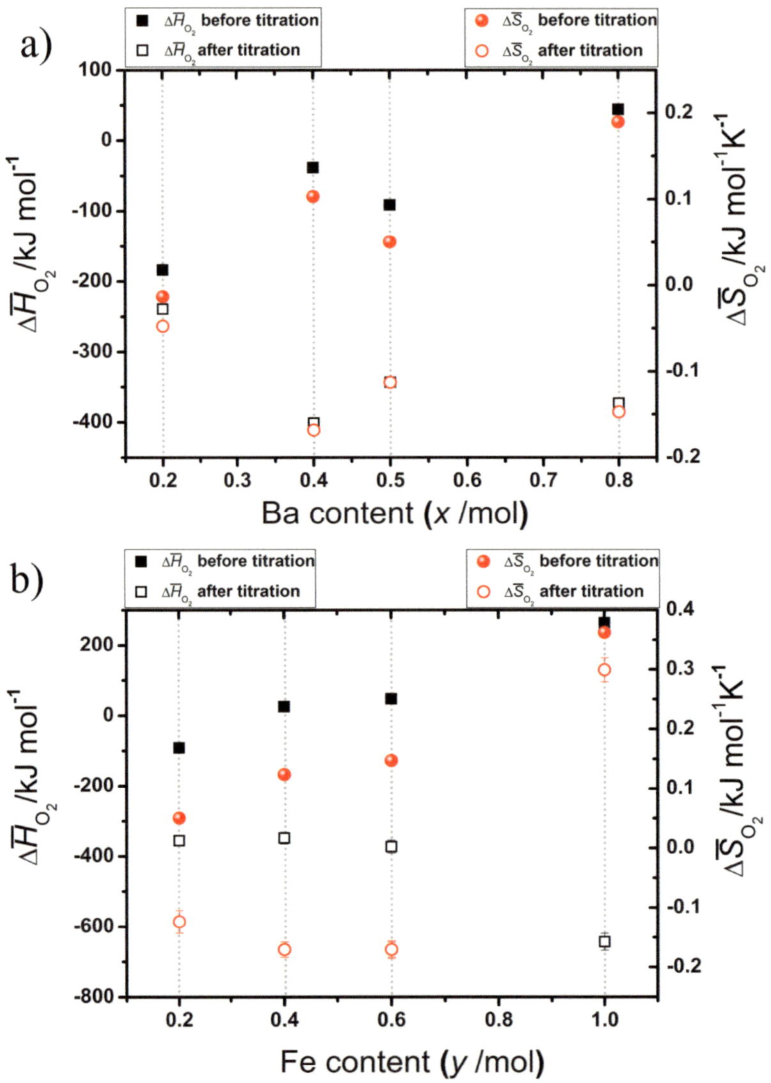

Figure 9.
$\Delta\bar{H}_{O_2}$ and $\Delta\bar{S}_{O_2}$ as a function of a) Ba and b) Fe content and illustrating the effect of the oxygen non-stoichiometry at temperatures above 1123 K.

random distribution of oxygen vacancies within the oxygen sublattice. The result is consistent with the morphological investigation as well as with the high stability of cubic perovskite phase evaluated for Sr- and Fe- rich compositions [20, 46].

The thermodynamics of solid solutions containing a mixture of cobalt and iron on the B-site is complex. Further details and measurements of the energy and the entropy of oxygen incorporation into BSCF at different values of non-stoichiometry δ are necessary in order to make clear the effect of the stoichiometry change on the vacancy distribution within the oxygen sublattice.

4. Conclusions

The thermodynamic behavior of BSCF compounds with different Ba and Fe contents, using an electrochemical cell with yttria-stabilized zirconia as solid electrolyte was investigated. The EMF measurements performed in a wide temperature range (823–1273 K) and at pressures of 10^{-5}–10^{-6} Pa confirm the instability of the BSCF-perovskite phases at temperatures lower than 1123 K.

With the help of the thermodynamic data, the points of phase transformations associated with the charge compensation mechanism were highlighted, the results being important for the assessment of the long-term stability of such nonstoichiometric materials used as cathodes in IT-SOFCs. The temperature associated to the structural transformations decreases with the increase of Ba-content. The thermodynamic investigation evidenced that for the system $Ba_xSr_{1-x}Co_{0.8}Fe_{0.2}O_{3-\delta}$, low symmetry BSCF-perovskites (BSCF 5582, BSCF 6482 and BSCF 4682) are thermodynamically more stable than high symmetry BSCF-perovskites (BSCF 2882 and BSCF 8282). In the case of $Ba_{0.5}Sr_{0.5}Co_{1-y}Fe_yO_{3-\delta}$ with different iron-content, at temperatures ≥1023 K, thermodynamic stability increased following the order $y = 1 < y = 0.6 < y = 0.4 < y = 0.2$, the composition BSCF 5582 being the most stable in the series, while the composition with $y = 0.8$ exhibited a peculiar behavior.

The BSCF compounds exhibited a significant variation of the thermodynamic parameters with the oxygen non-stoichiometry change, this variation being highly dependent on temperature and dopant concentration. The thermodynamic data evidenced that after decreasing the oxygen stoichiometry with the same relative deviation of $\Delta\delta = 0.01$, the specimen BSCF 5501 was the most stable composition within 1123–1273 K temperature range. The results of the thermodynamic study of BSCF compounds, at temperatures of interest for the application in IT-SOFC help us to explain why the BSCF 5582 composition exhibits the highest ionic conductivity and the highest oxygen catalytic activity.

Knowing the specific thermodynamic quantities of BSCF compositions, it is possible to find new routes to modify the properties of these materials by suitable substitution and formation of oxygen vacancies in oxygen lattice. The evaluation of thermodynamic quantities is mandatory to understand the complex relationships between the defect chemistry and the material properties.

Author details

Florentina Maxim[1], Alina Botea-Petcu[1], Florina Teodorescu[1]*, Ludwig J. Gauckler[2] and Speranta Tanasescu[1]

1 "Ilie Murgulescu" Institute of Physical Chemistry of the Romanian Academy, Laboratory of Chemical Thermodynamics, Bucharest, Romania

2 Nonmetallic Inorganic Materials, Department of Materials, ETH Zürich, Zürich, Switzerland

*Address all correspondence to: fteodorescu@icf.ro

References

[1] Chen ZH, Ran R, Zhou W, Shao ZP, Liu SM. Assessment of $Ba_{0.5}Sr_{0.5}Co_{1-y}Fe_yO_{3-\delta}$ (y=0.0–1.0) for prospective application as cathode for IT-SOFCs or oxygen permeating membrane. Electrochim Acta. 2007;52(25):7343–51. DOI: 10.1016/j.electacta.2007.06.010

[2] McIntosh S, Vente J, Haije W, Blank D, Bouwmeester H. Structure and oxygen stoichiometry of $SrCo_{0.8}Fe_{0.2}O_{3-\delta}$ and $Ba_{0.5}Sr_{0.5}Co_{0.8}Fe_{0.2}O_{3-\delta}$. Solid State Ionics. 2006;177(19–25):1737–42. DOI: 10.1016/j.ssi.2006.03.041

[3] McIntosh S, Vente JF, Haije WG, Blank DHA, Bouwmeester HJM. Oxygen stoichiometry and chemical expansion of $Ba_{0.5}Sr_{0.5}Co_{0.8}Fe_{0.2}O_{3-\delta}$ measured by *in situ* neutron diffraction. Chem Mater. 2006;18(8):2187–93. DOI: 10.1021/cm052763x

[4] Vente JF, McIntosh S, Haije WG, Bouwmeester HJM. Properties and performance of $Ba_xSr_{1-x}Co_{0.8}Fe_{0.2}O_{3-\delta}$ materials for oxygen transport membranes. J Solid State Electr. 2006;10(8):581–8. DOI: 10.1007/s10008-006-0130-2

[5] Bucher E, Egger A, Ried P, Sitte W, Holtappels P. Oxygen nonstoichiometry and exchange kinetics of $Ba_{0.5}Sr_{0.5}Co_{0.8}Fe_{0.2}O_{3-\delta}$. Solid State Ionics. 2008;179(21–26):1032–5. DOI: 10.1016/j.ssi.2008.01.089

[6] Sadykov VA, Sadovskaya EM, Uvarov NF. Methods of isotopic relaxations for estimation of oxygen diffusion coefficients in solid electrolytes and materials with mixed ionic-electronic conductivity. Russ. J. Electrochem. 2015;51(5):458–67. DOI: 10.1134/S1023193515050109

[7] Jung J-I, Misture ST, Edwards DD. The electronic conductivity of $Ba_{0.5}Sr_{0.5}Co_xFe_{1-x}O_{3-\delta}$ (BSCF: x = 0 ~ 1.0) under different oxygen partial pressures. J Electroceram. 2009;24(4): 261–9. DOI: 10.1007/s10832-009-9567-x

[8] Shao Z, Haile SM. A high-performance cathode for the next generation of solid-oxide fuel cells. Materials for Sustainable Energy. p. 255–8. DOI: 10.1038/nature02863

[9] Sahini MG, Tolchard JR, Wiik K, Grande T. High temperature X-ray diffraction and thermo-gravimetrical analysis of the cubic perovskite $Ba_{0.5}Sr_{0.5}Co_{0.8}Fe_{0.2}O_{3-\delta}$ under different atmospheres. Dalton T. 2015;44(23): 10875–81. DOI: 10.1039/C4DT03963G

[10] Waindich A, Möbius A, Müller M. Corrosion of $Ba_{1-x}Sr_xCo_{1-y}Fe_yO_{3-\delta}$ and $La_{0.3}Ba_{0.7}Co_{0.2}Fe_{0.8}O_{3-\delta}$ materials for oxygen separating membranes under Oxycoal conditions. J Membrane Sci. 2009;337(1):182–7. DOI: 10.1016/j.memsci.2009.03.041

[11] Yáng Z, Harvey AS, Gauckler LJ. Influence of CO_2 on $Ba_{0.2}Sr_{0.8}Co_{0.8}Fe_{0.2}O_{3-\delta}$ at elevated temperatures. Scripta Mater. 2009;61(11): 1083–6. DOI: 10.1016/j.scriptamat.2009.08.039

[12] Saša Z, Toni I, Sebastien V, Dijana J, J. GL. The changes of $Ba_{0.5}Sr_{0.5}Co_{0.8}Fe_{0.2}O_{3-\delta}$ perovskite oxide on heating in oxygen and carbon dioxide atmospheres. J Serb Chem Soc. 2014;79(9):13. DOI: 10.2298/JSC131024018Z

[13] Tanasescu S, Yang Z, Martynczuk J, Varazashvili V, Maxim F, Teodorescu F, Botea A, Totir N, Gauckler LJ. Effects of A-site composition and oxygen nonstoichiometry on the thermodynamic stability of compounds in the Ba-Sr-Co-Fe-O system. J Solid State Chem. 2013;200:354–62. DOI: 10.1016/j.jssc.2013.01.030

[14] Wang F, Nakamura T, Yashiro K, Mizusaki J, Amezawa K. The crystal

structure, oxygen nonstoichiometry and chemical stability of $Ba_{0.5}Sr_{0.5}Co_{0.8}Fe_{0.2}O_{3-\delta}$ (BSCF). Phys Chem Chem Phys. 2014;16(16):7307–14. DOI: 10.1039/C3CP54810D

[15] Wang HH, Tablet S, Yang WS, Caro R. *In situ* high temperature X-ray diffraction studies of mixed ionic and electronic conducting perovskite-type membranes. Mater Lett. 2005;59(28): 3750–5. DOI: 10.1016/j.matlet.2005.06.067

[16] Zhou W, Ran R, Shao Z. Progress in understanding and development of $Ba_{0.5}Sr_{0.5}Co_{0.8}Fe_{0.2}O_{3-\delta}$-based cathodes for intermediate-temperature solid-oxide fuel cells: A review. J Power Sources. 2009;192(2):231–46. DOI: 10.1016/j.jpowsour.2009.02.069

[17] Ovenstone J, Jung JI, White JS, Edwards DD, Misture ST. Phase stability of BSCF in low oxygen partial pressures. J Solid State Chem. 2008;181(3):576–86. DOI: 10.1016/j.jssc.2008.01.010

[18] Jung JI, Misture ST, Edwards DD. Seebeck coefficient and electrical conductivity of BSCF ($Ba_{0.5}Sr_{0.5}Co_xFe_{1-x}O_{3-\delta}$, $0 \leq x \leq 0.8$) as a function of temperature and partial oxygen pressure. Solid State Ionics. 2012;206: 50–6. DOI: 10.1016/j.ssi.2011.09.023

[19] Mueller DN, De Souza RA, Yoo H-I, Martin M. Phase stability and oxygen nonstoichiometry of highly oxygen-deficient perovskite-type oxides: A case study of (Ba,Sr)(Co,Fe)$O_{3-\delta}$. Chem Mater. 2012;24(2):269–74. DOI: 10.1021/cm2033004

[20] Yang Z, Harvey AS, Infortuna A, Gauckler LJ. Phase relations in the Ba-Sr-Co-Fe-O system at 1273 K in air. J Appl Crystallogr. 2009;42:153–60. DOI: 10.1107/S0021889809002040

[21] Yang Z, Harvey AS, Infortuna A, Schoonman J, Gauckler LJ. Electrical conductivity and defect chemistry of $Ba_xSr_{1-x}Co_yFe_{1-y}O_{3-\delta}$ perovskites. J Solid State Electr. 2011;15(2):277–84. DOI: 10.1007/s10008-010-1208-4

[22] Frank L, Hovorka M, Mikmeková Š, Mikmeková E, Müllerová I, Pokorná Z. Scanning electron microscopy with samples in an electric field. Materials. 2012;5(12):2731–56. DOI: 10.3390/ma5122731

[23] Tanasescu S, Totir ND, Marchidan DI. Thermodynamic data of the perovskite-type $LaMnO_{3\pm x}$ and $La_{0.7}Sr_{0.3}MnO_{3\pm x}$ by a solid-state electrochemical technique. Electrochim Acta. 1998;43(12–13):1675–81. DOI: 10.1016/S0013-4686(97)00311-3

[24] Tanasescu S, Totir ND, Marchidan DI, Turcanu A. The influence of compositional variables on the thermodynamic properties of lanthanum strontium ferrite manganites and lanthanum strontium manganites. Mater Res Bull. 1997;32(7):915–23. DOI: 10.1016/S0025-5408(97)00054-8

[25] Tanasescu S, Totir ND, Marchidan DI. Thermodynamic properties of the $SrFeO_{2.5}$ and $SrMnO_{2.5}$ brownmillerite-like compounds by means of EMF-measurements. Solid State Ionics. 2000;134(3–4):265–70. DOI: 10.1016/S0167-2738(00)00731-1

[26] Toprak MS, Darab M, Syvertsen GE, Muhammed M. Synthesis of nanostructured BSCF by oxalate co-precipitation – As potential cathode material for solid oxide fuels cells. Int J Hydrogen Energ. 2010;35(17):9448–54. DOI: 10.1016/j.ijhydene.2010.03.121

[27] Deganello F, Liotta L, Marcì G, Fabbri E, Traversa E. Strontium and iron-doped barium cobaltite prepared by solution combustion synthesis: exploring a mixed-fuel approach for tailored intermediate temperature solid oxide fuel cell cathode materials. Mater Renew Sustain Energy. 2013;2(1):1–14. DOI: 10.1007/s40243-013-0008-z

[28] Giuliano A, Carpanese MP, Clematis D, Boaro M, Pappacena A, Deganello F, et al. Infiltration, Overpotential and Ageing Effects on Cathodes for Solid Oxide Fuel Cells: $La_{0.6}Sr_{0.4}Co_{0.2}Fe_{0.8}O_{3-\delta}$ *versus* $Ba_{0.5}Sr_{0.5}Co_{0.8}Fe_{0.2}O_{3-\delta}$. J Electrochem Soc. 2017;164(10):F3114-F22. DOI: : 10.1149/2.0161710jes

[29] Zhao HL, Shen W, Zhu ZM, Li X, Wang ZF. Preparation and properties of $Ba_xSr_{1-x}Co_yFe_{1-y}O_{3-\delta}$ cathode material for intermediate temperature solid oxide fuel cells. J Power Sources. 2008; 182(2):503–9. DOI: 10.1016/j.jpowsour.2008.04.046

[30] Berenov A, Atkinson A, Kilner J, Ananyev M, Eremin V, Porotnikova N, et al. Oxygen tracer diffusion and surface exchange kinetics in $Ba_{0.5}Sr_{0.5}Co_{0.8}Fe_{0.2}O_{3-\delta}$. Solid State Ionics. 2014;268:102–9. DOI: 10.1016/j.ssi.2014.09.031

[31] Jung J-I, Jeong HY, Kim MG, Nam G, Park J, Cho J. Fabrication of $Ba_{0.5}Sr_{0.5}Co_{0.8}Fe_{0.2}O_{3-\delta}$ Catalysts with Enhanced Electrochemical Performance by Removing an Inherent Heterogeneous Surface Film Layer. Adv Mater. 2015;27(2):266–71. DOI: 10.1002/adma.201403897

[32] Müller P, Stormer H, Dieterle L, Niedrig C, Ivers-Tiffee E, Gerthsen D. Decomposition pathway of cubic $Ba_{0.5}Sr_{0.5}Co_{0.8}Fe_{0.2}O_{3-\delta}$ between 700 and 1000 °C analyzed by electron microscopic techniques. Solid State Ionics. 2012;206:57–66. DOI: 10.1016/j.ssi.2011.10.013

[33] Yang Z, Martynczuk J, Efimov K, Harvey AS, Infortuna A, Kocher P, et al. Oxygen-vacancy-related structural phase transition of $Ba_{0.8}Sr_{0.2}Co_{0.8}Fe_{0.2}O_{3-\delta}$. Chem Mater. 2011;23(13):3169–75. DOI: 10.1021/cm200373r

[34] Efimov K, Xu Q, Feldhoff A. Transmission electron microscopy study of $Ba_{0.5}Sr_{0.5}Co_{0.8}Fe_{0.2}O_{3-\delta}$perovskite decomposition at intermediate temperatures. Chem Mater. 2010;22(21):5866–75. DOI: 10.1021/cm101745v

[35] Harvey AS, Litterst FJ, Yang Z, Rupp JLM, Infortuna A, Gauckler LJ. Oxidation states of Co and Fe in $Ba_{1-x}Sr_xCo_{1-y}Fe_yO_{3-\delta}$ (x, y=0.2–0.8) and oxygen desorption in the temperature range 300–1273 K. Phys Chem Chem Phys. 2009;11(17):3090–8. DOI: 10.1039/B819414A

[36] Kuklja MM, Mastrikov YA, Jansang B, Kotomin EA. First principles calculations of (Ba,Sr)(Co,Fe)$O_{3-\delta}$ structural stability. Solid State Ionics. 2013;230:21–6. DOI: 10.1016/j.ssi.2012.08.022

[37] Merkle R, Mastrikov YA, Kotomin EA, Kuklja MM, Maier J. First principles calculations of oxygen vacancy formation and migration in $Ba_{1-x}Sr_xCo_{1-y}Fe_yO_{3-\delta}$ perovskites. J Electrochem Soc. 2012;159(2):B219-B26. DOI: 10.1149/2.077202jes

[38] Jung J-I, Misture ST, Edwards DD. Oxygen stoichiometry, electrical conductivity, and thermopower measurements of BSCF ($Ba_{0.5}Sr_{0.5}Co_xFe_{1-x}O_{3-\delta}$, $0 \leq x \leq 0.8$) in air. Solid State Ionics. 2010;181(27–28): 1287–93. DOi: 10.1016/j.ssi.2010.06.033

[39] Magnone E, Miyayama M, Traversa E. Structural properties and electrochemical characteristics of $Ba_{0.5}Sr_{0.5}Co_{1-x}Fe_xO_{3-\delta}$ phases in different atmospheres. J Electrochem Soc. 2009;156(9):B1059-B66. DOI: 10.1149/1.3158745

[40] Magnone E, Miyayama M, Traversa E. Some structural considerations on the perovskite-type $A_{1-y}Sr_yCo_{1-x}Fe_xO_{3-\delta}$ solid solution series. Cryst Res Technol. 2010;45(4):355–64. DOI: 10.1002/crat.200900704

[41] Celorrio V, Tiwari D, Fermin DJ. Composition-dependent reactivity of $Ba_{0.5}Sr_{0.5}Co_xFe_{1-x}O_{3-\delta}$ toward the

Oxygen Reduction Reaction. J Phys Chem C. 2016;120(39):22291–7. DOI: 10.1021/acs.jpcc.6b04781

[42] Lankhorst MHR, Bouwmeester HJM, Verweij H. High-temperature coulometric titration of $La_{1-x}Sr_xCoO_{3-\delta}$: Evidence for the effect of electronic band structure on nonstoichiometry behavior. J Solid State Chem. 1997;133(2):555–67. DOI: 10.1006/jssc.1997.7531

[43] Carter S, Selcuk A, Chater RJ, Kajda J, Kilner JA, Steele BCH. Oxygen transport in selected nonstoichiometric perovskite-structure oxides. Solid State Ionics. 1992;53–56, Part 1(0):597–605. DOI: 10.1016/0167-2738(92)90435-R

[44] Tanasescu S, Totir ND, Marchidan DI. Thermodynamic properties of some perovskite type oxides used as SOFC cathode materials. Solid State Ionics. 1999;119(1–4):311–5. DOI: 10.1016/S0167-2738(98)00520-7

[45] Fisher CAJ, Yoshiya M, Iwamoto Y, Ishii J, Asanuma M, Yabuta K. Oxide ion diffusion in perovskite-structured $Ba_{1-x}Sr_xCo_{1-y}Fe_yO_{2.5}$: A molecular dynamics study. Solid State Ionics. 2007;177(39–40):3425–31. DOI: 10.1016/j.ssi.2006.03.060

[46] Wessel C, Lumey MW, Dronskowski R. First-principles electronic-structure calculations on the stability and oxygen conductivity in $Ba_{0.5}Sr_{0.5}Co_{0.8}Fe_{0.2}O_{3-\delta}$. J Membrane Sci. 2011;366(1–2):92–6. DOI: 10.1016/j.memsci.2010.09.046

[47] Girdauskaite E, Ullmann H, Al Daroukh M, Vashook V, Bulow M, Guth U. Oxygen stoichiometry, unit cell volume, and thermodynamic quantities of perovskite-type oxides. J Solid State Electr. 2007;11(4):469–77. DOI: 10.1007/s10008-006-0175-2

[48] Attaoua M, Beriala S, Omari M. Defect chemistry and physical properties of $Ln_{0.5}Sr_{0.5}FeO_3$ (Ln: La, Pr). J Saudi Chem Soc. 2012;16(1):91–5. DOI: 10.1016/j.jscs.2010.11.004

[49] Adler SB. Factors governing oxygen reduction in solid oxide fuel cell cathodes. Chem Rev. 2004;104(10): 4791–844. DOI: 10.1021/cr020724o

[50] Adler S, Russek S, Reimer J, Fendorf M, Stacy A, Huang QZ, et al. Local-structure and oxide-ion motion in defective perovskites. Solid State Ionics. 1994;68(3–4):193–211. DOI: 10.1016/0167-2738(94)90177-5

[51] Hagenmuller P, Pouchard M, Grenier JC. Nonstoichiometry in the Perovskite-type oxides - an evolution from the classical Schottky-Wagner model to the recent high-Tc Superconductors. Solid State Ionics. 1990;43:7–18. DOI: 10.1016/0167-2738(90)90464-3

Chapter 6

Stoichiometric and Nonstoichiometric Compounds

Paras Dubey and Netram Kaurav

Abstract

This chapter gives a general overview of synthesis and recent development of nickel oxide as a nonstoichiometric compound. We establish the synthesis chemistry of nickel oxide as a nonstoichiometric material, and hence successively introduce definitions and classifications of nonstoichiometric compounds as well as their point defects. The samples of nonstoichiometric nickel oxide are synthesized by thermal decomposition method. The nonstoichiometry of samples was then studied chemically by iodometric titration, and the results are further corroborated by excess oxygen obtained from the thermo-gravimetric analysis (TGA). X-ray diffraction (XRD) and Fourier transformed infrared (FTIR) techniques are used to analyze structural phase of nonstoichiometric nickel oxide. The change in oxidation state of nickel was studied by X-ray photoelectron spectroscopy (XPS) analysis. The shift in antiferromagnetic ordering and transition temperature due to nonstoichiometry is studied by magnetic and specific heat capacity analysis.

Keywords: nonstoichiometry, defect study, nickel oxide, thermo-gravimetrical analysis, iodometric titrations

1. Introduction

In solid-state chemistry, the study of compound has been expanded to the crystal structure level. The law of definite proportions, the law of constant composition, and the law of conservation of mass state that a chemical compound always contains exactly the same proportion of elements by mass. All chemical compounds obeying these laws are called as stoichiometric compounds. While nonstoichiometric compounds are the chemical compounds deviated from stoichiometry, namely their elemental composition cannot be represented by a ratio of well-defined natural numbers, and therefore violate the law of definite proportions, hence a nonstoichiometric compound is a type of special solid-state compound with definite structure and thermodynamic characteristics, which differs from its stoichiometric counterpart and a mixture. Due to defect structure in a continuous manner, these compounds are different with stoichiometric compounds. Beside some unusual information on solid-state chemistry, stability, and dynamics can be explore through their structure and characteristic. These nonstoichiometric compounds exhibit different properties such as conductivity, magnetism, catalytic nature, color, and other unique solid-state properties, which have important technological applications.

Experimental and calculated results demonstrate that the migration of anion and cation vacancies and the formation of insulating vacancy clusters near the vicinity of the interface are the fundamental switching mechanism [1]. The enhancement in photoconduction under below-band gap light illumination is attributed to the transition from defect levels (e.g., oxygen vacancies) [2]. The unique properties are determined by their solid phase composition, thermodynamic properties, and crystal defect structure. These properties depend upon the structure and phase of the crystal; hence, the unit cell parameter plays an important role in nonstoichiometric compound, which varies in a definite manner. The microscopic structural and macroscopic thermodynamical properties also vary with phase and unit cell parameters. The X-ray diffraction study reveals that as the composition of the nonstoichiometric compound changes, the cell parameter and other property continuously changes, which is different from crystalline stoichiometric compound. Hence, X-ray diffraction and chemical analysis are important tools to investigate the nonstoichiometry [2]. The lattice defect can be analyzed by their different properties [2]. Hence, nonstoichiometry plays very important role in the solid compounds of transition metals and becoming a challenge and new opportunity in chemistry. Therefore, study of the nonstoichiometric compounds has become an interesting field in solid-state chemistry and defect chemistry.

There are significant structural changes that are seen in nonstoichiometric compounds. The phase of matter changes as external conditions such as pressure, temperature, or chemical composition are varied. Due to sudden change in these external thermodynamical conditions, the crystalline structure also changes. Thus, an abrupt change in thermodynamical conditions, such as specific volume, entropy, or specific heat, also changes the derivatives of free energy. Therefore, these parameters can be used as a helpful indicator of phase transitions. The abrupt changes can identify them as driven by a pseudo phase transition that is a result of a sharp switch of predominant defect species [3]. The defect that is introduced by sudden change in conditions, changes the physical properties of a material. It is quite different from the abrupt variation in physical properties that arises from symmetry or ordering change in a conventional phase transition. Therefore, according to the conventional classification, a defective state of a material cannot be labeled as a distinct physical phase compared with the perfect one.

In this chapter, we tried to understand the effect of change in stoichiometry of nickel oxide compound. We have prepared the samples of nickel oxide with thermal decomposition method. The samples were analyzed by different characterized methods and found to nonstoichiometric. The effect of nonstiochiometry on oxidation state, bonding between oxygen and nickel, magnetic property, specific heat, and structural property were studied.

2. Classification and point defect formation in nonstoichiometric compounds

The compounds with uniform physical phase in which unit cell parameters vary with its composition in a continuous manner are nonstoichiometric. The free energy of such compounds is function of composition and temperature of the system. The chemical potential depends on the composition of solid, and due to microscopic and macroscopic characteristics, the structural and thermodynamical property changes. Different parameters of crystallography cell, obtained by X-ray diffraction analysis, change due to nonstoichiometry. This parameter continuously varies with its composition [3].

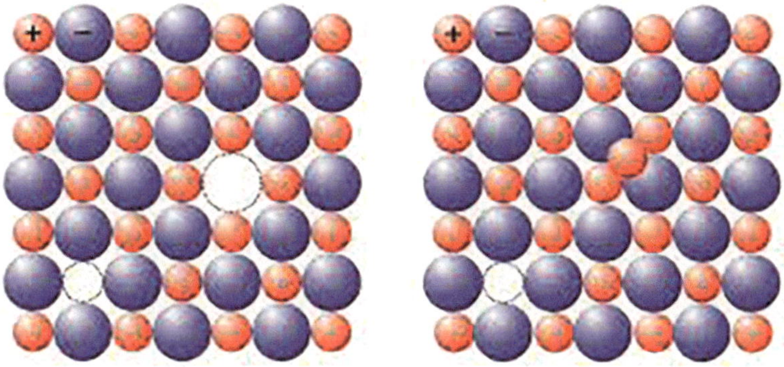

Figure 1.
Schottky and Frenkel defects.

Element composition is the main criteria to classify the nonstoichiometric compounds. These compounds exist only in the condensed state. Transition metal oxides are generally nonstoichiometric, some nitrides and sulfides also comes under such category [4–6]. Most nonstoichiometric compounds have compositions that are close to those of stoichiometric compounds and can be expressed by formulas such as WO_{3-x}, $Co_{1-x}O$, $Zn_{1+x}O$, and $Ni_{1-x}O$, in which x is a positive quantity much smaller than 1. In the first case, an oxygen vacancy would be formed, and in some cases, there is metallic deficiency that is seen. Due to deviation from stoichiometric composition, it would result in the formation of some lattice imperfections. Hence, in some nonstoichiometric compounds, there are anionic vacancies and in some, there are cationic vacancies. These vacancies are formed due to defects in the lattice structures of crystalline substances, such as the absence of ions from sites that would normally be occupied. The crystallographic point defects are the main defects in which interstitial atoms and vacancies resulting from excess or deficiency of a component element. Point defects are an important cause of formation of the nonstoichiometric compounds. There are two main point defects in nonstoichiometric compounds namely Frenkel defect and Schottky defect. The Frenkel defect explains a defect in the molecule, where an atom or ion (normally, the cation) leaves its own lattice site vacant and instead occupies a normally vacant site, while the Schottky defect forms when vacancies are created when oppositely charged ions leave their lattice sites. Formation of theses vacancies in stoichiometric system helps in maintaining an overall neutral charge in the ionic solid. These created vacancies are filled by the movement of surrounding atoms, due to which new vacancies are formed. The formation of defects in the crystal will lead to a decrease in the density of the crystal or metal. This can be understood by **Figure 1**.

3. Nonstoichiometry in nickel oxide

Nickel oxide is an antiferromagnetic [7] and Mott-Hubbard insulator [7]. Nickel oxide (NiO) has been among some of the transition metal oxide, which is most thoroughly studied by the researchers. It can be easily synthesized and is very low in preparation cost, the nature of this material is such that it has low toxicity and it is environmental friendly. These properties of nickel oxide attracted considerable interest of the researchers. Due to its versatility, nickel oxide is useful in number of applications, such as transparent conductive film [8], chemical sensors [9], and resistive random access memory [10]. The positive electrode in batteries [11] and in quantum dot light emitting devices as a hole transport layer [12] are some of the very latest and important applications of nickel oxide material. The condition of synthesis of the material plays very important role in the applications. If the condition of sample preparation changes, then the oxygen content of the sample also changes, and NiO of different stoichiometry ($Ni_{1-\delta}O$) can be obtained. In such samples, the composition ratios between nickel and oxygen are not exactly 1:1. Because of excess oxygen and vacancies on Ni site, nickel oxide thus becomes a p-type metal-deficit semiconductor [8]. The findings of investigations done by authors suggest that Ni vacancy is the most dominant point defect present in the system, rather than oxygen interstitial [13, 14]. The distribution of vacancy over a particular volume is an important issue. Due to difference in the nature of surface and bulk, the defect formation and hence distribution of vacancy will be different. The charge transport in NiO, as in other transition metal oxides, is of the thermally activated hopping type, and electron holes are localized at cation lattice sites. The nature of defect clusters and complexes in metal-deficit NiO, as well as the role of specific impurities, influence remarkably the different characteristic properties of nonstoichiometric NiO. This specific vacancy distribution has its strong effect on the overall electrical [15], optical [16], and thermal [17] properties, depending strongly on its stoichiometry.

The main reason for changes in properties could be the excess oxygen present in the samples. This excess oxygen also changes the oxidation state of nickel, which produces a vacancy at the metal side to produce cation vacant nickel oxide. As the temperature of sintering of precursor increases, the excess of oxygen decreases. Due to increase in the excess of oxygen, densities of nickel oxide and activation energy of electrical conductivity decrease, but the lattice parameter of such samples is unaffected [18]. The diffusion of vacancies to the crystal surface is the main reason for the decrease in vacancies [19].

NiO adopts the NaCl structure, with octahedral Ni(II) and O^{2-} sites. The conceptually simple structure is commonly known as the rock salt structure. Like many other binary metal oxides, the nonstoichiometry is accompanied by a color change, with the stoichiometrically correct NiO being green and the nonstoichiometric NiO being black. Bulk material of NiO has a cubical crystal structure, and it shows antiferromagnetic behavior below Neel temperature, and above it, the structure is transformed into a cubic one with paramagnetic behavior [20]. The change in the particle size is attributed to the excess oxygen present in nonstoichiometric nickel oxide. Scientist suggested that due to the change in oxygen content of the sample, the complexity of the disordered arrangement, volume distribution, and random orientation of the magnetization vector changes; hence, thermal, optical, electrical, magnetic, and many other properties of sample change. Due to the change in the particle size of antiferromagnetic materials sintered at different temperatures, the surface to volume ratio becomes sufficiently large. Because of their large surface area, the existing uncompensated spins give a nonzero net magnetic moment.

4. Synthesis of nonstoichiometric nickel oxide

Preparation of compound plays an important role in its characteristic properties. Nonstoichiometric compounds, whose composition and structure are known, are very important as they have unique optical, electrical, magnetic, thermal, and mechanical properties. The selection of precursor for design and tailoring the materials of different stoichiometry for desirable performance is of great interest. By means of effective and precise control of the composition, defect, and structure, the nonstoichiometry can be created. Depending on the conditions of preparation, the sintering temperature, in particular, NiO samples of various surface areas, colors, and degrees of nonstoichiometry can be prepared. Nickel oxide can easily be prepared via several methods, including chemical route, evaporation [12], sputtering [8], chemical deposition [21], oxidation of nickel [22], sol-gel method [23], and thermal decomposition [24]. Thermal decomposition method is a typical powder preparation method, which is simple, low-cost, and fast endothermic process.

Nonstoichiometric nickel oxide was obtained by thermal decomposition of nickel nitrate hexahydrate [25]. Typically, about 5 g $Ni(NO_3)_2.6H_2O$ was decomposed thermally in open air for 3 hours at 400°C to produce nickel oxide sample with a particular content of oxygen ($Ni_{1-\delta}O$). The product thus obtained was pure and contains no other mixture. This exothermic reaction is noncatalytic and does not require any other substance to initiate the reaction. The output material obtained by breaking down the precursor is nickel oxide and gaseous parts, which are escaped from the system freely. The decomposition mechanism of this process takes the minimum time-temperature condition. This sample was denoted as NiO400. Similarly, seven other samples were prepared at 500, 600, 700, 800, 900, 1000, and 1100°C. They were denoted NiO500, NiO600, NiO700, NiO800, NiO900, NiO1000, and NiO1100, respectively.

5. Characterization of nonstoichiometric nickel oxide

The study of compositions, substance phase, and defect are main parameters to characterize nonstoichiometric compounds. The effect of nonstoichiometry on composition, structure, and other properties is studied while characterizing these types of solids. These properties highly depend on the defects in the crystal. In the present study, composition can be determined by means of chemical analysis by iodometric titration, which was corroborated by thermo-gravimetric analysis (TGA). Phase and structure can be examined through X-ray diffraction, XPS analyses, and relationship of the composition, structure, and properties can be detected using FTIR, magnetic measurement, and specific heat measurement. So, we can categorically confirm that the nickel oxide is a nonstoichiometric compound rather than a stoichiometric compound. Furthermore, this FTIR also confirms the bonding of oxygen with metal ions, and shift in the FTIR peaks indicates change in stoichiometry of the as prepared samples.

Oxidation sate along with chemical nature and binding state were analyzed by X-ray Photoelectron Spectroscopy (XPS). This method is a surface analytical method, hence all the properties that change with composition were studied by this method. The behavior of transition temperature with change in stoichiometry was studied by temperature dependence of magnetic susceptibility (c) and heat capacity (Cp).

5.1 Structure determination of nonstoichiometric nickel oxide

To confirm the phase structure of polycrystalline powder sample, X-ray diffraction technique is one of the most important methods. This method can be used for the determination of the phase structure and unit cell parameters. The XRD measurements of the nonstoichiometric nickel oxide samples are conducted with the help of Bruker D8 Advance X-ray diffractometer. Cu Kα radiation (0.154 nm) is used for the analysis, and measurements were done in the angle ranging 10–90°. Silicon strip technology is used to detect the scattered X-ray radiations.

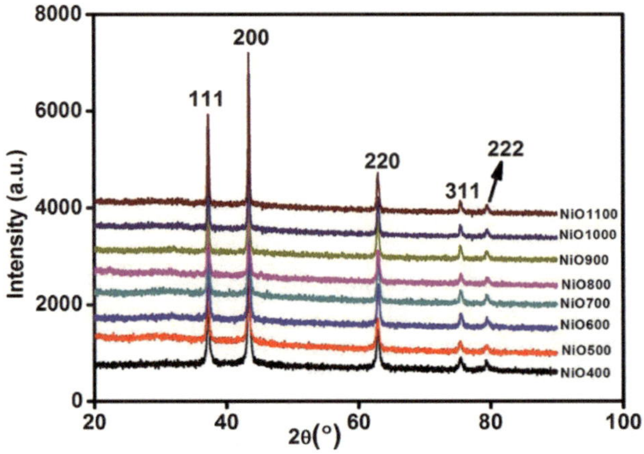

Figure 2.
X-ray diffraction patterns of different nonstoichiometric samples.

Intensity (%)	2θ	Sin^2 θ	$h^2 + k^2 + l^2$	C	(hkl)	a(Å)	ā(Å)
61.69	37.279	0.1022	3	0.034	(111)	4.1772	4.1775
100	43.472	0.1365	4	0.034	(200)	4.1727	
44.83	62.901	0.2722	8	0.034	(220)	4.1772	
15.68	75.416	0.3741	11	0.034	(311)	4.1772	

Table 1.
Crystallographic data using analytical analysis for NiO samples.

XRD analysis was employed to investigate the crystallinity and purity of the solid product obtained by thermal conversion in air of the polycrystalline compounds. **Figure 2** shows XRD patterns of $Ni_{1-\delta}O$ samples, NiO400, NiO500, NiO700, NiO1100 and NiO600, NiO800, NiO900, and NiO1000. XRD peaks match well with the standard XRD of with no other impurity peaks, showing that these samples were of a single phase in nature in each of them. The XRD data show the presence of the characteristic peaks for NiO at 2θ in the range of 10–90°, in accordance with JCPDF File 47-1049. The lattice parameters were obtained from the following relationship analytically: $\sin 2\theta = C\ (h^2 + k^2 + l^2)$, where $C = \lambda^2/4a^2$. The results of the X-ray structural analysis are given in **Table 1**.

On comparing peaks obtained in X-ray analysis with standard JCPDF File 47-1049, we conclude that the precursor salt is transformed to NiO. Hence, data from analytical and structural analysis corroborate with the experimental data. This indicates that the NiO obtained by thermal conversion in air of the poly-nuclear coordination compounds is face-centered-cubic phase, also known as the bunsenite structure (lattice constant a of cubic unit cell, 0.4177 nm). Phase of the NiO prepared by thermal decomposition method is stable, as no other peaks were seen in the structural analysis.

Further, Sherrer formula: $L = k\lambda/\beta\cos(\theta)$ is used to calculate the mean crystallite size of the NiO sample where L is the crystallite size, k is the Sherrer constant, usually taken as 0.89, λ is the wavelength of the X-ray radiation (0.154056 nm for Cu Kα), and β is the full width at half maximum (FWHM) of diffraction peak measured at 2θ.

The deduced values of crystallite size of the analyzed NiO samples are reported in **Table 2**, which indicates that as the sintering temperature of precursor increases the crystallite size of NiO changes.

Compounds	a (Å)	V (Å^3)	Rf factor	Bregg's R-factor	R_p	R_{wp}	R_e	χ^2	*L* (nm)
NiO (400°C)	4.176	72.82	2.39	3.94	24	16.7	14.9	1.26	240
NiO (500°C)	4.174	72.74	4.44	4.33	29	19.6	16.7	1.38	290
NiO (700°C)	4.172	72.64	4.04	4.51	22	15.8	13.3	1.16	350
NiO (1100°C)	4170	27.58	20.4	31.7	56	37.4	22.3	2.67	460

Table 2.
Rietveld parameters of different NiO samples.

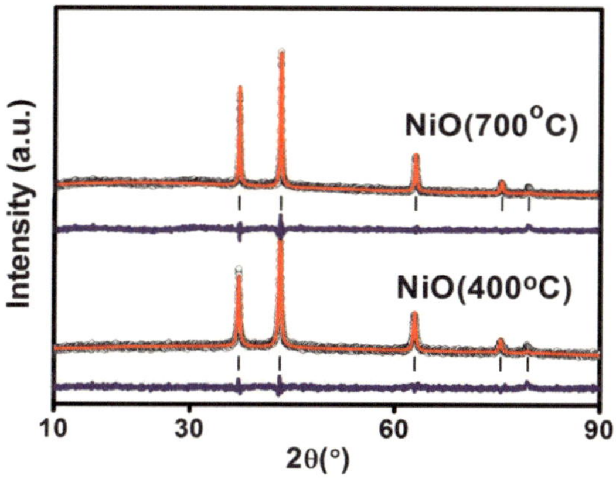

Figure 3.
Rietveld profile fit for NiO400 and NiO700 samples.

XRD patterns were profile-refined using the Full-Prof software package (http://www-llb.cea.fr/fullweb/). **Figure 3** gives the representative Rietveld profile fit for NiO400 and NiO700 samples along with different patterns obtained by using Fm3m (225) space group. The XRD analysis indicates that samples of nickel oxide have only one face-centered-cubic (fcc) phase. **Table 2** gives the parameters of samples found in the characterization. XRD study reveals all the peaks of NiO that are present, and the calculation indicates unit cell volume changes with stoichiometry.

5.2 Stoichiometry and defect analysis of nickel oxide

In nonstoichiometric compounds, there exist defects in the lattice structures of crystalline substances. As there are no absolute defect-free crystals at T > 0 K, hence these defects are related to nonstoichiometry. To determine oxygen content, that is, change in vacancy of nickel oxide formed by thermal decomposition method, the redox iodometric titration is the suitable method. For iodometric titration, standardized sodium thiosulfate and potassium iodide solution were used as a titrant and titrand, and starch solution is used as an endpoint indicator. The result obtained from iodometric titration was matched by the results of thermo-gravimetric analysis (TGA). The TGA system with the top of the line METTLER TOLEDO ultra-micro balance with unique built-in calibration weights ensures an accuracy of 0.1 mg was used.

For iodometric titrations, 0.0025 g sample was dissolved in KI and HCl solution. The concentration of KI was 0.1 mol^{-1} and concentration of HCl was ca. 0.1 mol^{-1}, titration was carried out using sodium thiosulfate (2.023×10^{-3} mol^{-1}) as a titrant. The starch solution was added prior to the end-point being reached as it acts as indicator. **Table 1** shows the results of excess oxygen in NiO samples obtained by iodometric titration. The oxygen content of the samples indicates that the ratio of oxygen concentration to nickel concentration is greater than 1 for those samples, which were prepared below 700°C, and as the temperature increases above 700°C, the O/Ni ratio becomes nearly equal to 1.

The result of iodometric titration indicates that, on sintering, the precursor below 700°C nickel oxide having different oxygen contents can be prepared. As the sintering temperature is raised above 700°C, the excess oxygen present in the samples evolved, and the sample becomes stoichiometric. It can also be inferred from the above results that the oxidation state of nickel ions is also affected by sintering temperature. This titration also gives the amount of Ni^{3+} present in the samples, and hence the percentage of excess oxygen can also be determined. This can be obtained by considering that two Ni^{3+} ions correspond to three ions of O^{2-}.

The excess oxygen content of the samples was also calculated by TGA. The TGA analysis was performed in the inert atmosphere, in which the change in weight of a sample was analyzed with respect to temperature. TGA characterizes the materials according to its compositions. For our samples, by obtaining the weight loss with respect to increase in temperature, the weight of the excess oxygen is determined directly. **Figure 4a** gives the information about the thermogravimetric (TG) curve and its differential curve for NiO400 sample. Between 200 and 300°C, the sudden change in the mass of the sample is observed, and this indicates the amount of excess oxygen released in the heating process. This change in weight can also be associated with the desorption of oxygen from sample, which confirms nonstoichiometry. The other peak seen in the curve can be associated by the

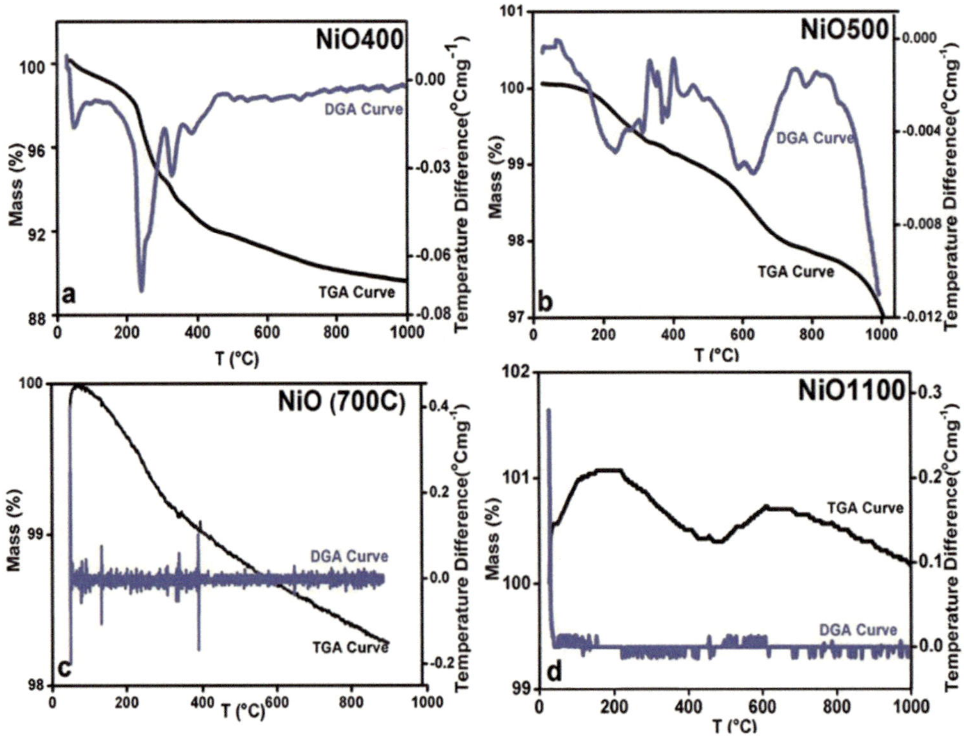

Figure 4.
TGA curves of nonstoichiometric $Ni_{1-\delta}O$ samples for (a) NiO400, (b) NiO500, (c) NiO700, and (d) NiO1100; heating rate of 5°C per min. The curves indicate the change in oxygen content of different samples as the sintering temperature increases.

evaporation of the physically bound water. Hence, this change in the weight does not have any contribution to the estimation of excess oxygen.

Unlike NiO400 TG and DGA curves for three more samples, that is, NiO500, NiO700, and NiO1100 were plotted and shown in **Figure 4b–d**. It is clear from the graphs that the loss in weight decreases as the sintering temperature for sample preparation increases. The change in mass for NiO500 sample is less than that of NiO400 sample and is negligible for the sample prepared above 700°C. The change in weight of samples directly indicates about excess oxygen present in the sample, which can be directly obtained from above TG curves and is shown in **Table 3**. Comparing the calculated excess oxygen in these nonstoichiometric samples by two different methods, it is understood that both results are similar to each other.

We have also prepared another sample of nickel oxide at 400°C, and the main difference is that this sample was prepared in the presence of oxygen. This sample is synthesized to understand the essential feature of excess oxygen in these nonstoichiometric samples. **Figure 5** shows the TG and DGA curves of this sample. The curves shown in **Figure 4a** and in **Figure 5** are almost similar. Hence, the samples of NiO400 prepared in air and in oxygen atmosphere are identical to each other. The weight of excess oxygen of sample prepared in presence of oxygen is slightly less as compared to NiO400 sample, which was prepared in air. The

Samples	Excess oxygen in iodometric titration (%)	Excess oxygen in thermogravimetric analysis (%)
NiO400	7.2	7.8 (6.4)*
NiO500	2.5	2.8
NiO600	1.4	1.6
NiO700	1.1	1.2
NiO800	0	0
NiO900	0	0
NiO1000	0	0
NiO1100	0	0

**% excess oxygen calculated for NiO400 sample prepared in presence of oxygen.*

Table 3.
Excess oxygen calculated from iodometric titration and TGA.

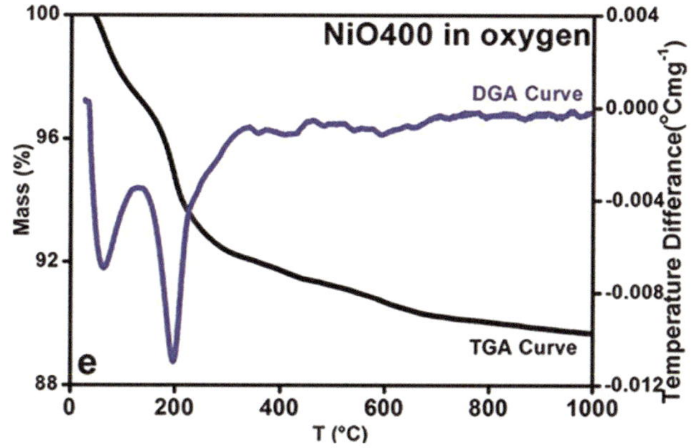

Figure 5.
TGA curve of NiO400 prepared in oxygen atmosphere.

difference in loss of weight percentage is calculated, and the values are reported in **Table 3**. Hence from iodometric and TGA analysis, it can be understood that due to rearrangement of atoms at high temperature, the defects of atoms get heals up and the sample that was nonstoichiometric at low temperature becomes stoichiometric. This is due to change in oxidation state with respect to change in temperature of sintering of precursor.

5.3 FTIR study of nonstoichiometric nickel oxide

FTIR spectroscopic study of $Ni_{1-\delta}O$ samples provides valuable information about the phase composition and the way in which oxygen is bonded to metal ions. **Figure 6a–d** show infrared (IR) transmission spectra of NiO400, NiO500, NiO700, and NiO1100 samples having different stoichiometry in the range between 400 and 4000 cm^{-1}. We observed a prominent peak in spectrum between 440 and 460 cm^{-1}. The slight shift in this peak for different samples is observed. This shift in the peak is noticed for all those samples that were sintered below 700°C. The reason of such peak shift can be attributed to stoichiometry of the samples. The observation of such peak in the long wavelength region, analogous to previous reports [26], could be assigned to the Ni–O stretching vibration mode, and the shifting is an indication of the nonstoichiometry present in these samples. In fact, in this long wavelength transverse optical mode, in which the sublattice of Ni^{2+} ions moves 180° opposite to the sublattice of O^{2-} ions for bulk NiO has been reported to lie between 390 and 405 cm^{-1} [27].

Some more peaks at 1032, 1383, and 1612 cm^{-1} in nonstoichiometric samples sintered below 700°C are seen prominently, which indicates the presence of hydroxide ions, nitrate ions, and some organic compounds. It is instructive to mention that the thermal analysis evidently identifies that the $Ni(NO_3)_2.6H_2O$ was decomposed completely to NiO at temperatures higher than 600°C [28]. The

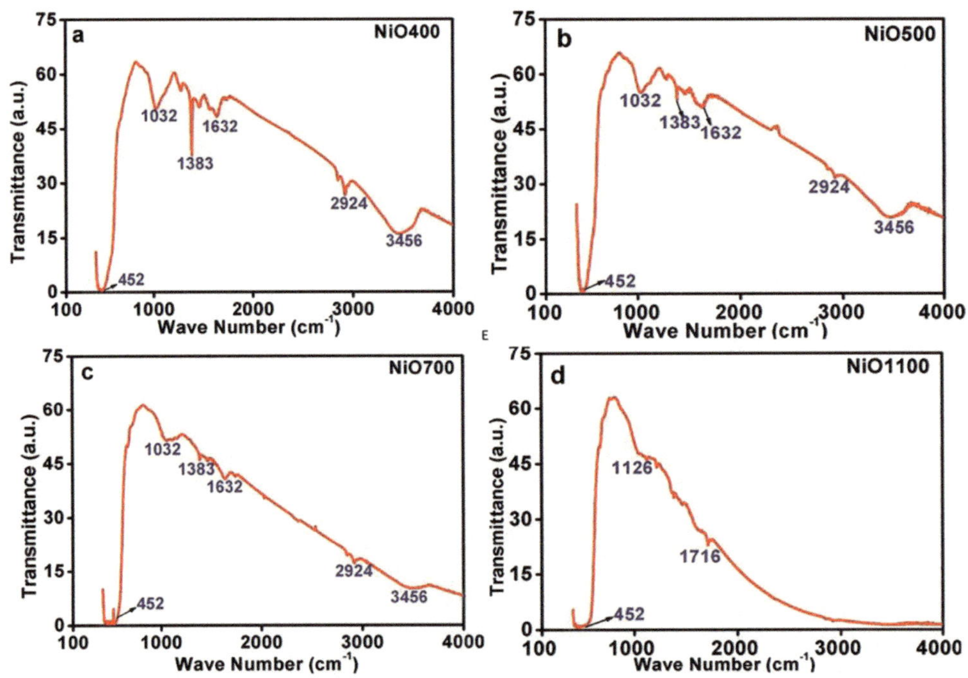

Figure 6.
FTIR of nonstoichiometric Ni1_dO samples for (a) NiO400, (b) NiO500, (c) NiO700, and (d) NiO1100.

samples that are sintered till 600°C may contain some organic molecules. Further, some of the bands found to be disappeared as the decomposition temperature increases as TGA data indict that at higher sintering temperature excess oxygen decreases [28]. Some carbon impurities are also present in the spectra, indicated by 2924 and 3456 cm^{-1} peaks in the spectrum. Hence, FTIR data shown in **Figure 6a–d**, TGA data shown in **Figure 4a–d**, and iodometric analysis data are corroborating each other and indicate that the samples are nonstoichiometric in nature. Analyzing the data obtained by the vacancy model, the concentration of hole in nickel oxide indicates that there is deficiency of metal in $Ni_{1-\delta}O$. Cationic vacancies of nickel in NiO are formed at nickel side due to the presence of excess oxygen. These vacancies can be ionized to create Ni^{3+} ions via the following reaction:

$$2Ni_{Ni}^{x} + \frac{1}{2}O_2 \rightarrow 2Ni_{Ni}^{\cdot} + O_O^{x} + V_{Ni}'' \tag{1}$$

one ionized nickel vacancy (V_{Ni}'') and two Ni^{3+} ions ($Ni_{Ni}^{\cdot}$) will produce when two Ni^{2+} ions (Ni_{Ni}^{x}) will react with oxygen to produce in the NiO crystal. The holes are donated by each created Ni^{3+} ion, which also alters the conductive NiO. The results of iodometric titration, TGA, and FTIR analyses indicate that nonstoichiometric nickel oxide is nickel-deficient. Hence, nonstoichiometric nickel oxide behaves as p-type semiconductor because according to Eq. (1), we argue that an increase in number of nickel hole concentration increases as Ni^{3+} ions increase.

The majority of defects in nonstoichiometric nickel oxide is electron holes, which creates the vacancies. The vacancy model crated by holes is confirmed by a Seebeck coefficient [29], electrical conductivity [30], and other measurements. The kinetic measurements of the rate of oxidation of nickel metal to nickel oxide are also given by vacancy model [31]. The physical and chemical properties of nonstoichiometric nickel oxide are altered by the defect processes, and hence excess oxygen becomes prominent role player in such crystals.

By calculating inter atomic potentials, the energies of defect formation, migration, and substitution can be calculated by the atomistic simulation [32]. In the present case, the potential describing interionic interactions is represented by ionic pair-wise potentials of the form Eq. (2) Here, the first term represents long-range Coulomb, the second term corresponds to Hafemeister and Flygare form of short-range repulsive energies [33], and Van der Waals multipole are represented by third and fourth terms, respectively.

$$\phi(r) = \sum_{ij} \frac{Z_m e^2}{r_{ij}} + \sum_{ij} b\beta_{ij} \exp\left(\frac{r_i + r_j - r_{ij}}{\rho}\right) + \sum_{ij} c_{ij} r_{ij}^{-6} + \sum_{ij} d_{ij} r_{ij}^{-8}. \tag{2}$$

The symbols: c_{ij} and d_{ij} are the Van der Waals coefficients, and β_{ij} is the Pauling coefficient, respectively. Z_m is the modified ionic charge and parametrically includes the Coulomb screening effect, while b and r are short-range parameters. Thus, the effective interionic potential contains only three free parameters (Z_m, b, and r), which can be determined from the crystal properties [34]. The short-range potential parameters assigned to each ion-ion interaction were derived by empirical fitting to observed structural properties. In the context of the Columbic term, integral ionic charges are presumed, that is, 2^+ for Ni and 2^- for O, which enables a straight forward definition of hole states as Ni^{3+} or O^-. The deduced potential parameters are listed in **Table 4** for all the samples. It is clear from the calculated parameters as the sintering temperature of precursor increases, the nature of ordering in systems in which ions occupy sites on a face-centered-cubic (fcc) lattice

Samples	b (10^{-12} erg)	P (Å)	F (eV)
NiO400	39.51	0.388	39.592
NiO500	39.46	0.388	39.602
NiO600	39.40	0.388	39.612
NiO700	39.38	0.388	39.623
NiO800	39.40	0.388	39.612
NiO900	39.41	0.388	39.612
NiO1000	39.57	0.388	39.518
NiO1100	39.41	0.388	39.612

Table 4.
Interatomic potential parameter of nickel oxide sintered at different temperatures as discussed in Eq. (2).

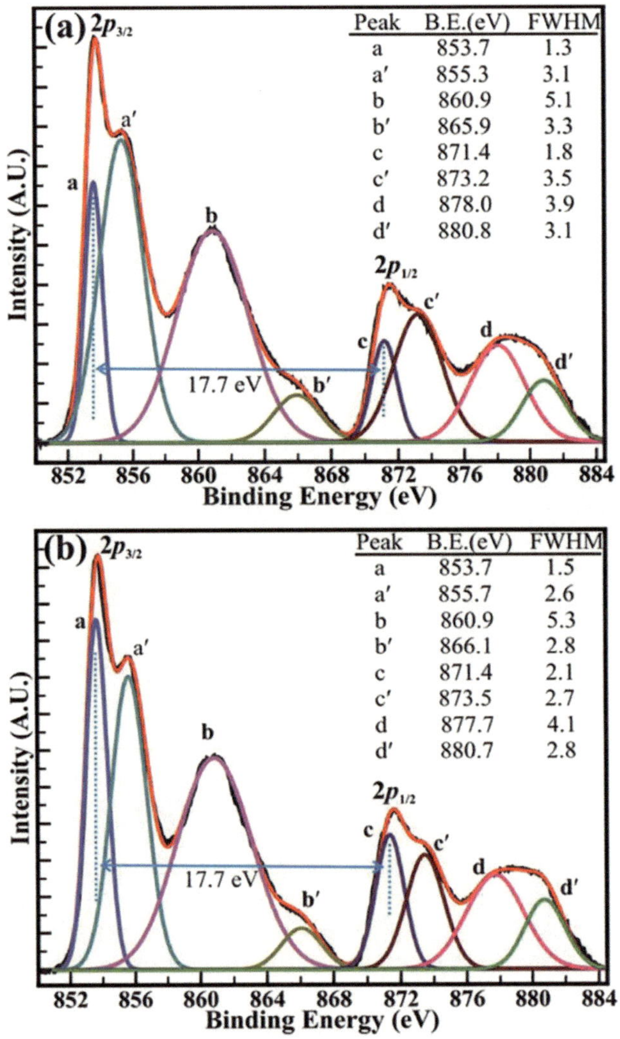

Figure 7.
High-resolution XPS spectra of the Ni (2p) core levels of the Ni oxides decomposed at a temperature of (a) 500° C and (b) 700°C. The XPS spectra were decomposed using Voigt peak function fittings.

changes and the nearest-neighbor (NN) and next-nearest neighbor (NNN) exchange interactions also get altered. This happens in such a manner that the structure of NiO undergoes a week cubic-to-rhombohedral distortion as a result of the magnetostriction effect in the presence of excess oxygen in the samples [35]. The NN and NNN exchange interactions and the antiferromagnetic (AFM) structure of NiO are altered due to the presence of excess oxygen in the samples. Since the radius of Ni^{3+} ions is smaller than that of Ni^{2+}. The Ni^{3+}-O^{2-} bond distance will be short for those samples whose oxygen content is higher. In due course, interatomic potential parameters and bond length change as the sintering temperature changes, the magnetic ordering transition temperature is expected to change as we will discuss in later sections.

5.4 XPS studies of nonstoichiometric nickel oxide

The chemical properties and stoichiometry of nickel oxide were studied by XPS analysis. For NiO500 and NiO700, XPS spectra of Ni (2p) core level are shown in **Figure 7a** and **b**, while the XPS spectra for NiO400 and NiO1100 are shown in **Figure 8**.

In both the spectra, distinct peaks for different binding energies can be observed clearly along with the double peak features. These double peak features represented for the Ni (2p) core levels are observed in all samples. Voigt peak fitting function is used to analyze the double peak features of $Ni(2p_{3/2})$ and $Ni(2p_{1/2})$. These peaks were fitted within the Shirley background. The binding energy of 853.7 ($\pm$0.2), 855.5 ($\pm$0.2), 860.6 ($\pm$0.2), 865.9 ($\pm$0.2), 871.4 ($\pm$0.2), 873.3 ($\pm$0.2), 877.9 ($\pm$0.2), and 880.7 ($\pm$0.2) eV represented by a, a′, b, b′, c, c′, d, and d′ are associated with all eight peaks, which are perfectly fit. The peaks marked as a, a′, c, and c′, represent core levels of $Ni^{2+}(2p_{3/2})$, $Ni^{3+}(2p_{3/2})$, $Ni^{2+}(2p_{1/2})$, and $Ni^{3+}(2p_{1/2})$, respectively. The decomposed shake-up satellite peaks (marked as b, b′, d, and d′) were observed at ~7.1($\pm$0.2) or 10.2($\pm$0.2) eV and ~6.3($\pm$0.3) or ~7.1($\pm$0.2) eV higher in binding energy than that of $Ni^{2+}(2p_{3/2})$, $Ni^{3+}(2p_{3/2})$, $Ni^{2+}(2p_{1/2})$, and $Ni^{3+}(2p_{1/2})$ peaks, respectively. The magnetic chemical state of Ni^{2+} and Ni^{3+} ions can be associated with the double peak features of Ni(2p). The consecutive shake-up satellite peaks also give information of the same [36]. The positions of the XPS peaks obtained in different nonstoichiometric samples are analogs to previous studies [37]. The O(1 s) XPS spectra of the samples obtained were similar to each other. **Figure 8** shows the O(1 s) spectra Voigt peak fitting function within the Shirley background. The binding energy of 529.3 and 531.1 eV with FWHM of 1.2 and 1.7 eV are clearly observed in the figure for oxygen spectra. Hence, two peaks are clearly obtained for O(1 s) spectra. The lower binding energy peak observed at 529.3 eV corresponds to the O(1 s) core level of O^{2-} anions associated with Ni-O chemical bonding. However, higher binding energy peak observed at 531.1 represents the surface contamination or presence of hydroxyl (−OH) groups [37].

The close analysis of decomposed XPS spectrum revealed that the intensity of the peaks assigned to core levels of $Ni^{3+}(2p_{3/2})$ and $Ni^{3+}(2p_{1/2})$ is larger than that of core levels $Ni^{2+}(2p_{3/2})$ and $Ni^{2+}(2p_{1/2})$ at decomposition temperature of 400 and 500°C. However, the intensity of the peaks of $Ni^{2+}(2p_{3/2})$ and $Ni^{2+}(2p_{1/2})$ core levels has increased than that of the peaks of $Ni^{3+}(2p_{3/2})$ and $Ni^{3+}(2p_{1/2})$ core levels after the decomposition temperature of 700°C and is continued to increase for the temperature of 1100°C. The intensity ratio obtained for the peaks of the $Ni^{2+}(2p_{3/2})$ and $Ni^{3+}(2p_{3/2})$ core levels (i.e., Ni^{2+}/Ni^{3+}) has increased from 0.85 ($\pm$0.03) to 1.23 ($\pm$0.03) with an increase in the temperature from 400 to 1100°C.

It is evident that for the samples prepared at 500 and 700°C have more prominent variation of intensity, and hence the Ni^{2+} and Ni^{3+} are more easily bifurcated at

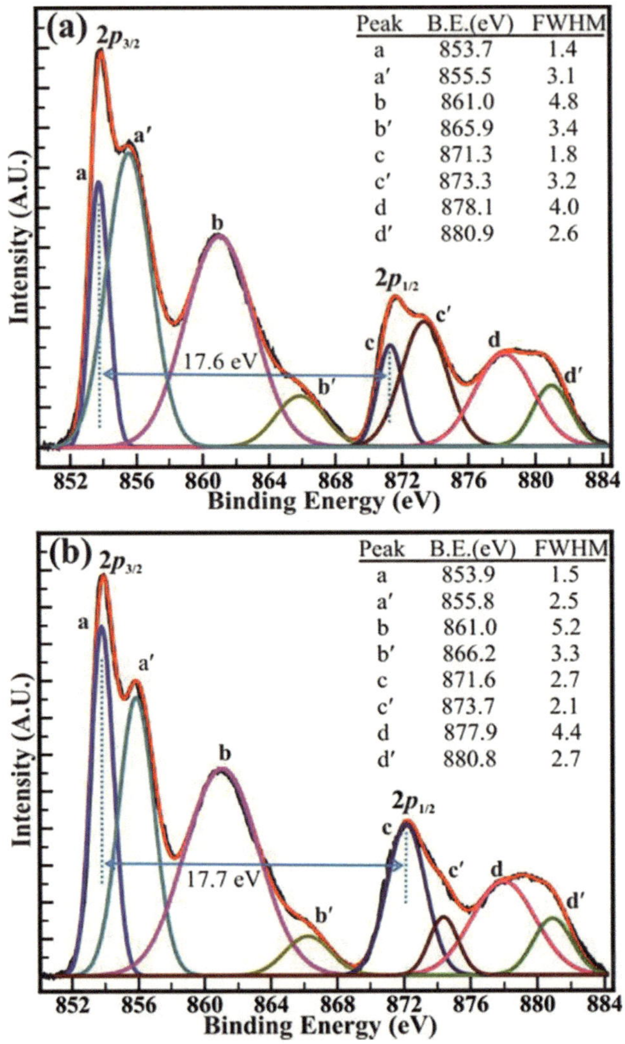

Figure 8.
High-resolution XPS spectra of the Ni (2p) core levels of the Ni oxides decomposed at a temperature of (a) 400° C and (b) 700°C. The XPS spectra were decomposed using Voigt peak function fittings.

these temperatures because the conversion of Ni^{3+} ions into Ni^{2+} ions is fast. And the binding energy difference (ΔE) of 17.7 ($\pm$0.1) eV between the $Ni(2p_{3/2})$ and Ni ($2p_{1/2}$) peaks is very close to that of 17.8 eV for oxidized Ni and significantly larger than that of 17.2 eV for metallic Ni [38]. Hence, all the nickel is converted into oxide form, and there is no nickel present in the metal form. This can be confirmed by XPS investigation that the nonstoichiometric NiO400 contains Ni^{3+} ions in higher quantity as compared to NiO1000. The XPS analysis also corroborates with our TGA analysis for nonstoichiometric nickel oxide. Hence, the oxygen content of the samples can be controlled by controlling the temperature of preparation of samples. Hence, nonstoichiometry in nickel oxide can be created by controlling temperature of preparation.

5.5 Magnetic properties measurements of nonstoichiometric nickel oxide

The magnetic characterization of nonstoichiometric samples was performed to obtain the magnetic susceptibility (χ). These measurements were done for selected

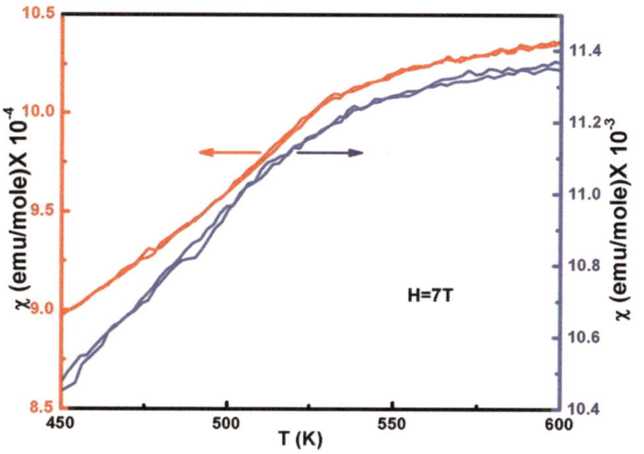

Figure 9.
Magnetization for ZFC and FC curves of nonstoichiometric $Ni_{1-\delta}O$ samples for NiO400 and NiO1100 in 7 T applied field as a function of temperature.

samples of NiO400 and NiO1100. Temperature-dependent magnetic susceptibility (χ) is shown in **Figure 9**. This measurement was performed at an applied magnetic field of 7 Tesla. Both zero-field-cooled (ZFC) and field-cooled (FC) conditions were employed to measure the susceptibility (χ) of nonstoichiometric nickel oxide samples. We observed a reversible behavior with negligible hysteresis of χ for different stoichiometry. It is interesting to describe that sample shows paramagnetic (PM) to antiferromagnetic (AFM) transition. This transition is observed at the Néel temperature $T_{N.}$ But the transition temperature at which transition takes place is different in both the samples. The values are 480 and 530 K for NiO400 and NiO1100 samples, respectively. T_N and the absolute value of χ for NiO400 sample are lower than that of NiO1100 sample. However, PM to AFM transition width appears to be broader for the NiO1100 sample. This can be understood as T_N changes with stoichiometry, which could be attributed to the effect of the partial destruction of Ni^{2+}–O–Ni^{3+} exchange interaction network because of the reduction of the oxygen vacancies and the weakening of Ni^{2+}–O–Ni^{3+} interaction arising from the decrease of the bandwidth of, for example, electrons due to the change in Ni–O bond length and Ni–O–Ni bond angle [39].

5.6 Specific heat measurement

Figure 10 shows the change in specific heat with respect to temperature of nonstoichiometric NiO. The irregularity or the change in specific heat for different samples is found in the vicinity of $T_{N.}$ This shows that AFM ordering of these samples is affected by oxygen content of the samples. A shift in anomaly is probed as the stoichiometry of sample changes. Change in heat capacity of different samples, due to stoichiometry, also alters the transition temperature. Hence, as the sintering temperature increases, T_N will change and finally attain a fixed value. It is observed from the heat capacity graph that the transition temperature for NiO500, NiO700, and NiO1100 samples are 510, 519, and 525 K, respectively. It is understood that due to excess oxygen in the different ratios in NiO400 and NiO500 samples, change in specific heat anomaly is observed. Further, the observed transition temperature reaches to 525 K as reported for stoichiometric NiO [40]. An irregular behavior in Cp at T_N is due to Ni-spin ordering as suggested by Néel [41] wherein due to the thermal agitation spin-lattice of the particle could reverse coherently and randomly.

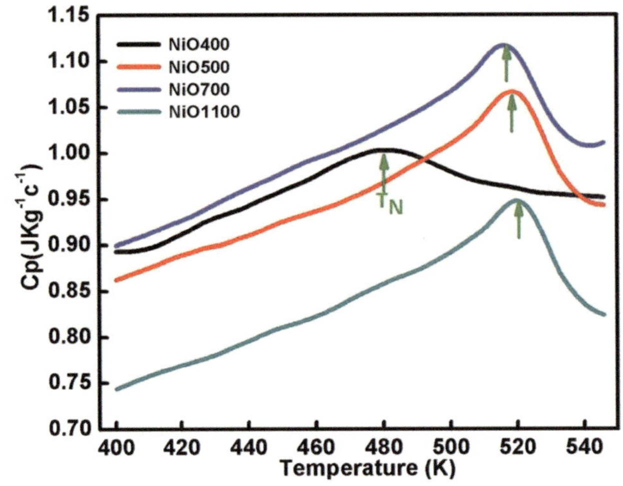

Figure 10.
The temperature variation of specific capacity of nickel oxide samples sintered at different temperatures. Curve shows shifting in TN due to different oxygen contents of the samples.

Consequently, the net moment of uncompensated surface spins would fluctuate accordingly and in turn, significant downshift the T_N in samples due to both the change of magnetic ions as well as their disorder. Probably due to change in excess oxygen, magnetic inhomogeneity in nonstoichiometric samples increases, and Néel temperatures eventually shift.

6. Conclusion

The samples of different oxygen contents of nickel oxide were synthesized by thermal decomposition method. The structure of these samples was characterized by X-ray diffraction spectroscopy, and full-proof refinement was used for indexing. The nonstoichiometry of these samples was recognized by iodometric titration and thermogravimetric analysis. Both these techniques give the excess oxygen content of samples, which was found to be similar. Different phases of nickel oxide were confirmed by the FTIR studies. The shift in FTIR pattern indicates about the nonstoichiometry of samples. The change in stoichiometry of sample is also responsible for the defect creation in the samples, as XPS results indicate that nickel vacancy can be created in samples with varying sintering temperatures. Those samples that were sintered at low temperature have high defect as compared to sample sintered at high temperature. The ZFC and FC curves are nearly identical to each other but the susceptibility changes with stoichiometry. The observed specific heat anomaly in the vicinity of T_N is associated with the magnetic ordering, indicating a gradual transformation between two magnetic phases, and the observed T_N shifted toward lower temperatures as excess oxygen content increases. The shifting of Néel temperature is presumably due to magnetic inhomogeneity arising from the excess oxygen in samples.

Acknowledgements

The authors would like to thank Dr. A. M. Awasti, Dr. Mukul Gupta, and Dr. V. Ganesan, Centre Director, UGC-DAE-CSR, Indore, India for providing heat capacity measurements and XRD facilities. Financial support from the Science and Engineering Research Board (SERB), the Department of Science and Technology (DST), New Delhi, and the Madhya Pradesh Council of Science and Technology (MPCST), Bhopal (NK) are gratefully acknowledged.

Conflict of interest

The authors declare that there are no conflicts of interest.

Author details

Paras Dubey[1] and Netram Kaurav[2*]

1 Department of Physics, Institute of Sciences, SAGE University, Indore, Madhya Pradesh, India

2 Department of Physics, Government Holkar Science College, Indore, Madhya Pradesh, India

*Address all correspondence to: kauravnetram@gmail.com; netramkaurav@yahoo.co.uk

References

[1] Kim YS, Kim J, Yoon MJ, Sohn CH, Lee SB, Lee D, et al. Impact of vacancy clusters on characteristic resistance change of nonstoichiometric strontium titanate nano-film. Applied Physics Letters. 2014;**104**: 013501-1-5. DOI: 10.1063/1.4955268

[2] Huang S, Ou G, Cheng J, Li H, Pan W. Ultrasensitive visible light photoresponse and electrical transportation properties of nonstoichiometric indium oxide nanowire arrays by electrospinning. Journal of Materials Chemistry C. 2013; **1**:6463-6470. DOI: 10.1039/C3TC31051E

[3] Yang Y, Liu XL, Yang Y, Xiao W, Li ZW, Xue DS, et al. Synthesis of nonstoichiometric zinc ferrite nanoparticles with extraordinary room temperature magnetism and their diverse applications. Journal of Materials Chemistry C. 2013;**1**: 2875-2885. DOI: 10.1039/C3TC00790A

[4] Fergus JW. Oxide materials for high temperature thermoelectric energy conversion. The Journal of the European Ceramic Society. 2012;**32**:525-540. DOI: 10.1016/j.jeurceramsoc.2011.10.007

[5] Lipatnikov VN, Zueva LV, Gusev AI, Kottar A. Disorder-order phase transformations and electrical resistivity of nonstoichiometric titanium carbide. Physics of the Solid State. 1998;**40**: 1211-1218. DOI: 10.1134/1.1130523

[6] Gusev AI. Order-disorder transformations and phase equilibria in strongly nonstoichiometric compounds. Uspekhi Fizicheskikh Nauk. 2003;**170**: 3-40. DOI: 10.3367/UFNr.0170.200001a.0003

[7] Sugiyma I, Shibata N, Wang Z, Kobayashi S, Yamamoto T, Ikuhara Y. Ferromagnetic dislocations in antiferromagnetic NiO. Nature Nanotechnology. 2013;**8**:266-270. DOI: 10.1038/nnano.2013.45

[8] Sato H, Minami T, Takata S, Yamada T. Transparent conducting p-type NiO thin films prepared by magnetron sputtering. Thin Solid Films. 1993;**236**:27-31. DOI: 10.1016/0040-6090(93)90636-4

[9] Hotovy I, Huran J, Spiess L, Hascik S, Rehacek V. Preparation of nickel oxide thin films for gas sensors applications. Sensors and Actuators, B: Chemical. 1999;**57**:147-152. DOI: 10.1016/S0925-4005(99)00077-5

[10] Shima H, Takano F, Akinaga H. Resistance switching in the metal deficient-type oxides: NiO and CoO. Applied Physics Letters. 2007;**91**: 012901. DOI: 10.1063/1.2753101

[11] Sun YK, Myung ST, Park BC, Prakash J, Belharouak I, Amine K. High-energy cathode material for long-life and safe lithium batteries. Nature. 2009; **8**:320-324. DOI: 10.1038/nmat2418

[12] Caruge J, Halpert JE, Bulovic V, Bawendi MG. NiO as an inorganic hole-transporting layer in quantum-dot light-emitting devices. Nano Letters. 2006;**6**: 2991-2994. DOI: 10.1021/nl0623208

[13] Ferrari AM, Pisani C, Cinquini F, Giordano L, Pacchioni G. Cationic and anionic vacancies on the NiO(100) surface: DFT+UDFT+U and hybrid functional density functional theory calculations. The Journal of Chemical Physics. 2007;**127**:174711-174716. DOI: 10.1063/1.2796154

[14] Nowotny J, Sadowski A. Application of microcalorimetry in studies of interaction between oxygen and NiO single crystals. Journal of the American Ceramic Society. 1979;**62**:241-253. DOI: 10.1007/BF00656843

[15] Kofstad P. Nonstoichiometry, Diffusion, and Electrical Conductivity in Binary Metal Oxides. New York: Wiley; 1972. DOI: 10.1007/BF01046721

[16] Conell RS, Corrigan DA, Powell BR. The electrochromic properties of sputtered nickel oxide films. Solar Energy Materials and Solar Cells. 1992; **25**:301-313. DOI: 10.1016/0927-0248 (92)90075-Z

[17] Patil PS, Kadam DS. Preparation and characterization of spray pyrolyzed nickel oxide (NiO) thin films. Applied Surface Science. 2002;**199**:211-221. DOI: 10.1016/S0169-4332(02)00839-5

[18] Iida Y, Kenjo S, Ozaki S. Non-stoichiometry of nickel oxide. Bulletin of the Chemical Society of Japan. 1960; **33**:1372-1375. DOI: 10.1246/bcsj.33.1372

[19] Chen HL, Lu YM, Hwang WS. Effect of film thickness on structural and electrical properties of sputter-deposited nickel oxide films. Materials Transactions. 2005;**46**:872-879. DOI: 10.2320/matertrans.46.872

[20] Kodama RH, Makhlouf SA, Berkowitz AE. Finite size effects in antiferromagnetic NiO nanoparticles. Physical Review Letters. 1997;**79**: 1393-1399. DOI: /10.1103/ PhysRevLett.79.1393

[21] Pramanik P, Bhattachraya S. A chemical method for the deposition of nickel oxide thin films. Journal of the Electrochemical Society. 1990;**137**:3869-3870. DOI: 10.1149/ 1.2086316

[22] Lampert CM, Omstead TR, Tu PC. Chemical and optical properties of electrochromic nickel oxide films. Solar Energy Materials. 1986;**14**:161. DOI: 10.1016/0165-1633(86)90043-2

[23] Surca A, Orel B, Pilhar B, Bwkovec P. Optical, spectroelectrochemical and structural properties of sol-gel derived Ni-oxide electrochromic film. The Journal of Electroanalytical Chemistry. 1996;**408**: 83-100. DOI: 10.1016/0022-0728(96) 04509-3

[24] Kim KS, Winograd N. X-ray photoelectron spectroscopic studies of nickel-oxygen surfaces using oxygen and argon ion-bombardment. Surface Science. 1974;**43**:625-643. DOI: 10.1016/ 0039-6028(74)90281-7

[25] Brockner W, Ehrhardt C, Gjikaj M. Thermal decomposition of nickel nitrate hexahydrate, $Ni(NO_3)_2{\cdot}6H_2O$, in comparison to $Co(NO_3)_2{\cdot}6H_2O$ and Ca $(NO_3)_2{\cdot}4H_2O$. Thermochimica Acta. 2007;**456**:64-68. DOI: 10.1016/j. tca.2007.01.031

[26] Yu C, Nazri G, Lampert CM. Spectroscopic and electrochemical studies of electrochromic hydrated nickel oxide films. Solar Energy Materials. 1987;**16**:1-17. DOI: 10.1016/ 0165-1633(87)90003-7

[27] Nakagawa I. Far-infrared reflection spectra, optical and dielectric constants, and lattice vibrations of some fluoride crystals. Bulletin of the Chemical Society of Japan. 1971;**44**:3014-3020. DOI: 10.1246/bcsj.44.3014

[28] Moore WJ. Seven Solid States. W. A. WA, USA: Benjamin, Inc; 1967

[29] Bosman AJ, Daal HJ, Van GF, Knuvers GF. Hall effect between 300° K and 1100° K in NiO. Physics Letters. 1965;**19**:372-373. DOI: 10.1016/ 0031-9163(65)90901-7

[30] Osburn CM, Vest RW. Defect structure and electrical properties of NiO—I. high temperature. Journal of Physics and Chemistry of Solids. 1971; **32**:1331-1342. DOI: 10.1016/S0022-3697 (71)80191-9

[31] Fueki K, Wegner JB. Studies of the oxidation of nickel in the temperature range of 900° to 1400°C. Journal of the Electrochemical Society. 1965;**112**: 384-388. DOI: 10.1149/1.2423553

[32] Catlow CRA. Computer simulation studies of transport in solids. Annual Review of Materials Science. 1986;**16**: 517-548. DOI: 10.1146/annurev.ms.16.080186.002505

[33] Hafemeister DW, Flygare WH. Outer-Shell overlap integrals as a function of distance for halogen—Halogen, halogen—Alkali, and alkali—Alkali ions in the alkali halide lattices. The Journal of Chemical Physics. 1965; **43**:795. DOI: 10.1063/1.1696846

[34] Kaurav N. High-pressure phase transformation and elastic behavior of XC (X = Si, Ge, Sn and Pt) compounds. Physica Scripta. 2013;**88**: 015604-015609. DOI: 10.1088/0031-8949/88/01/015604

[35] Mironova-Ulmane N, Skvortsova V, Kuzmin A. Magnetic ion exchange interactions in NiO-MgO solid solutions. Physics of the Solid State. 2005;**47**: 1516-1522. DOI: 10.1134/1.2014504

[36] Shaikh JS, Pawar RC, Devan RS, Ma YR, Salvi PP, Kolekar SS, et al. Synthesis and characterization of Ru doped CuO thin films for supercapacitor based on Bronsted acidic ionic liquid. Electrochimica Acta. 2011;**56**:2127-2134. DOI: 10.1016/j.electacta.2010.11.046

[37] Wagner CD, Riggs WM, Davis LE, Moulder JF, Muilenberg GE. Handbook of X-Ray Photoelectron Spectroscopy. Minnesota: Perkin-Elmer Corporation; 1979. DOI: 10.1002/sia.740030412

[38] Devan RS, Lin C-L, Gao S-Y, Cheng C-L, Liou Y, Ma Y-R. Room-temperature wide-range photoluminescence and semiconducting characteristics of two-dimensional pure metallic Zn nanoplates. RSC Advances. 2012;**2**:2123-2127. DOI: 10.1039/C2RA00972B

[39] Lenglet M, Hochu F, Durr J, Tuilier MH. Investigation of the chemical bonding in $3d^8$ nickel(II) charge transfer insulators (NiO, oxidic spinels) from ligand-field spectroscopy, Ni $2p$ XPS and X-ray absorption spectroscopy. Solid State Communications. 1997;**104**:793-798. DOI: 10.1016/S0038-1098(97)00273-1

[40] Hutchings MT, Samuelsen EJ. Measurement of spin-wave dispersion in NiO by inelastic neutron scattering and its relation to magnetic properties. Physical Review B. 1972;**6**:3447-3456. DOI: 10.1103/PhysRevB.6.3447

[41] N'eel L. Superposition de l'antiferromagnétisme et du superparamagnétisme dans un grain très fin. Comptes Rendus de l'Académie des Sciences. 1961;**252**:4075-4079